ζ Zeta

Decimals and Percents

Instruction Manual

By Steven P. Demme

1-888-854-MATH (6284) - www.MathUSee.com
Sales@MathUSee.com

Math·U·See

Building Understanding

Zeta Instruction Manual: Decimals and Percents ©2012 Math-U-See
Zeta Teacher Manual ©2004 Steven P. Demme
Published and distributed by Math-U-See

1·888·854·MATH (6284) or +1 717·283·1448

Sales@MathUSee.com

Lancaster, Pennsylvania USA

www.MathUSee.com

ISBN 978-1-60826-084-3

Printed in United States of America   1060-002-012   0113

1 2 3 4 5   09 10 11 ·12 13

Building Understanding in Teachers & Students to Nurture a Lifelong Love of Learning

At Math-U-See, our goal is to build understanding for all students.

We believe that education should be relevant, skill-based, and built on previous learning. Because students have a variety of learning styles, we believe education should be multi-sensory. While some memorization is necessary to learn math facts and formulas, students also must be able to apply this knowledge in real-life situations.

Math-U-See is proud to partner with teachers and parents as we use these principles of education to **build lifelong learners.**

CURRICULUM SEQUENCE

Calculus

PreCalculus
with Trigonometry

Algebra 2

Geometry

Algebra 1

Pre-Algebra

Zeta
Decimals and Percents

Epsilon
Fractions

Delta
Division

Gamma
Multiplication

Beta
Multiple-Digit Addition and Subtraction

Alpha
Single-Digit Addition and Subtraction

Primer
Introducing Math

Math-U-See is a complete, comprehensive K-12 math curriculum that uses manipulatives to illustrate and teach math concepts. We strive toward "Building Understanding" by using a multi-sensory, mastery-based approach suitable for all levels and learning styles. While each book concentrates on a specific theme, other math topics are introduced where appropriate. Subsequent books continuously review and integrate topics and concepts presented in previous levels.

Where to Start

Because Math-U-See is mastery-based, students may start at any level. We use the Greek alphabet to show the sequence of concepts taught rather than the grade level. Go to MathUSee.com for more placement help.

Each level builds on previously learned skills to prepare a solid foundation so the student is then ready to apply these concepts to algebra and other upper level courses.

Major concepts and skills:

- Expanding understanding of place value from positive powers of ten to include decimals
- Fluently adding, subtracting, multiplying, and dividing multiple digit decimals using place-value strategies
- Solving real-world problems with decimals and percents

Additional concepts and skills:

- Understanding and simplifying exponents
- Understanding negative numbers and representing them on the coordinate plane
- Using properties of operations to simplify and evaluate algebraic expressions
- Interpreting and graphing relationships between dependent and independent variables
- Understanding of plane geometry & geometric symbols
- Using ratio reasoning to solve problems
- Measuring statistical variability and distributions

Find more information and products at MathUSee.co

ζ Zeta

HOW TO USE

Five Minutes for Success

Welcome to *Zeta*. I believe you will have a positive experience with the unique Math-U-See approach to teaching math. These first few pages explain the essence of this methodology which has worked for thousands of students and teachers. I hope you will take five minutes and read through these steps carefully.

I am assuming your student has a thorough grasp of the four basic operations (addition, subtraction, multiplication, and division) and a mastery of fractions.

If you are using the program properly and still need additional help, you may visit us online at MathUSee.com or call us at 888-854-6284. –**Steve Demme**

The Goal of Math-U-See

The underlying assumption or premise of Math-U-See is that the reason we study math is to apply math in everyday situations. Our goal is to help produce confident problem solvers who enjoy the study of math. These are students who learn their math facts, rules, and formulas and are able to use this knowledge to solve word problems and real life applications. Therefore, the study of math is much more than simply committing to memory a list of facts. Our program includes memorization, but it also encompasses learning the underlying concepts of math that are critical to successful problem solving.

Support and Resources

Math-U-See has a number of resources to help you in the educational process.

Many of our customer service representatives have been with us for over 10 years. They are able to answer your questions, help you place your student in the appropriate level, and provide knowledgeable support throughout the school year.

Visit MathUSee.com to use our many online resources, find out when we will be in your neighborhood, and connect with us on social media.

More than Memorization

Many people confuse memorization with understanding. Once while I was teaching seven junior high students, I asked how many pieces they would each receive if there were fourteen pieces. The students' response was, "What do we do: add, subtract, multiply, or divide?" Knowing how to divide is important, understanding when to divide is equally important.

The Suggested 4-Step Math-U-See Approach

In order to train students to be confident problem solvers, here are the four steps that I suggest you use to get the most from the Math-U-See curriculum.

Step 1. Prepare for the Lesson
Step 2. Present and Explore the New Concept Together
Step 3. Student Practices for Mastery
Step 4. Student Progresses after Mastery

Step 1. Prepare for the Lesson

Watch the video lesson to learn the new concept and see how to demonstrate this concept with the manipulatives when applicable. Study the written explanations and examples in the instruction manual.

Step 2. Present and Explore the New Concept Together

Present the new concept to your student. Have the student watch the video lesson with you, if you think it would be helpful. The following should happen interactively.

a. **Build:** Use the manipulatives to demonstrate and model problems from the instruction manual. If you need more examples, use the appropriate lesson practice pages.

b. **Write:** Write down the step-by-step solutions as you work through the problems together, using manipulatives.

c. **Say:** Talk through the why of the math concept as you build and write.

Give as many opportunities for the student to "Build, Write, Say" as necessary until the student fully understands the new concept and can demonstrate it to you confidently. One of the joys of teaching is hearing a student say *"Now I get it!"* or *"Now I see it!"*

Step 3. Student Practices for Mastery

Using the lesson practice problems from the student workbook, have students practice the new concept until they understand it. It is one thing for students to watch someone else do a problem, it is quite another to do the same problem themselves. Together complete as many of the lesson practice pages as necessary (not all pages may be needed) until the student understands the new concept, demonstrating confident mastery of the skill. Remember, to demonstrate mastery, your student should be able to teach the concept back to you using the Build, Write, Say method. Give special attention to the word problems, which are designed to apply the concept being taught in the lesson.

Go to MathUSee.com to find review tools and other resources.

Step 4. Student Progresses after Mastery

Once mastery of the new concept is demonstrated, advance to the systematic review pages for that lesson. These worksheets review the new material as well as provide practice of the math concepts previously studied. If the student struggles, reteach these concepts to maintain mastery. If students quickly demonstrate mastery, they may not need to complete all of the systematic review pages.

In the 2012 student workbook, the last systematic review page for each lesson is followed by a page called "Application and Enrichment." These pages provide a way for students to review and use their math skills in a variety of different formats. Some of the Application and Enrichment pages introduce terms and ideas that a student may encounter on standardized tests. Mastery of these concepts is *not* necessary in order to move to the next level of Math-U-See. You may decide how useful these activity pages are for your particular student.

Now you are ready for the lesson tests. These were designed to be an assessment tool to help determine mastery, but they may also be used as extra worksheets. Your student will be ready for the next lesson only after demonstrating mastery of the new concept and maintaining mastery of concepts found in the systematic review worksheets.

Tell me, I forget. Show me, I understand. Let me do it, I remember.
–Ancient Proverb

To which Math-U-See adds, *"Let me teach it, and I will have achieved mastery!"*

Length of a Lesson

How long should a lesson take? This will vary from student to student and from topic to topic. You may spend a day on a new topic, or you may spend several days. There are so many factors that influence this process that it is impossible to predict the length of time from one lesson to another. I have spent three days on a lesson, and I have also invested three weeks in a lesson. This experience occurred in the same book with the same student. If you move from lesson to lesson too quickly without the student demonstrating mastery, the student will become overwhelmed and discouraged as he or she exposed to more new material without having learned previous topics. If you move too slowly, the student may become bored and lose interest in math. I believe that as you regularly spend time working along with the student, you will sense the right time to take the lesson test and progress through the book.

By following the four steps outlined above, you will have a much greater opportunity to succeed. Math must be taught sequentially, as it builds line upon line and precept upon precept on previously learned material. I hope you will try this methodology and move at the student's pace. As you do, I think you will be helping to create a confident problem solver who enjoys the study of math.

Exponents; Word Problem Tips

Multiplication is a shortcut for fast adding of the same number. To take this a step further, a shortcut for fast multiplying of the same number is raising to a power, which is represented with *exponents.* You can think of exponents in several ways. Picture a square that is 10 over and 10 up. It is 10 two ways (over and up), it is also a square or "ten squared." By definition, the exponent—in this case 2—stands for how many times 10 is used as a factor. The number 10 is used as a factor twice.

Another way to state it is "ten to the two power" or "ten to the power of two." Any number may be expressed as a square, such as "two squared" or "three squared." Look at figure 1 for examples. Notice the many ways to express the same idea.

Figure 1

5 over and 5 up
5 used as a factor 2 times
5 to the second power
5 squared
5 to the 2 power
5 raised to the power of 2

$5^2 = 5 \cdot 5 = 25$

3 over and 3 up
3 used as a factor 2 times
3 to the second power
3 squared
3 to the 2 power
3 raised to the power of 2

$3^2 = 3 \cdot 3 = 9$

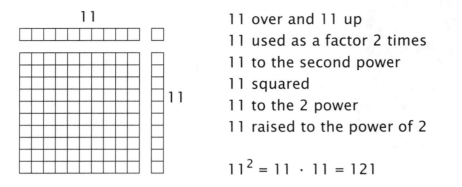

11 over and 11 up
11 used as a factor 2 times
11 to the second power
11 squared
11 to the 2 power
11 raised to the power of 2

$11^2 = 11 \cdot 11 = 121$

While it is difficult to show larger exponents with the manipulatives, you can have numbers raised to other powers than two. An example is $2 \times 2 \times 2 = 2^3$, which is 8. Another example is $3 \cdot 3 \cdot 3 \cdot 3 = 3^4$. It helps to say, "Three used as a factor four times is the same as 3^4, which is 81."

Example 1
$7^2 = 7 \cdot 7 = 49$

Example 2
$10 \times 10 \times 10 \times 10 \times 10 = 10^5 = 100,000$

When a number is raised to a power of three, for example 4^3, it may be read as "four to the third power," or "four cubed." You can make a figure that is a cube with dimensions 4 by 4 by 4. See figure 2.

Figure 2

4 over and 4 up and 4 deep
4 used as a factor 3 times
4 to the third power
4 cubed
4 to the 3 power
4 raised to the power of 3

$4^3 = 4 \cdot 4 \cdot 4 = 64$

Word Problem Tips

Parents often find it challenging to teach children how to solve word problems. Here are some suggestions for helping your student learn this important skill.

The first step is to realize that word problems require both reading and math comprehension. Don't expect a child to be able to solve a word problem if he does not thoroughly understand the math concepts involved. On the other hand, a student may have a math skill level that is stronger than his or her reading-comprehension skills. Below are a number of strategies to improve comprehension skills in the context of story problems. You should decide which ones work best for you and your child.

Strategies for word problems

1. Ignore numbers at first and read the story. It may help some students to read the question aloud. Every word problem tells a story. Before deciding what math operation is required, let the student retell the story in his own words. Who is involved? Are they receiving gifts, losing something, or dividing a treat?

2. Relate the story to real life, perhaps by using names of family members. For some students, this may make the problem more interesting and relevant.

3. Build, draw, or act out the story. Use the blocks or actual objects when practical. Especially in the lower levels, you may require the student to use the blocks for word problems even when the facts have been learned. Don't be afraid to use a little drama as well. The purpose is to make it as real and meaningful as possible.

4. Look for the common language used in a particular kind of problem. Pay close attention to the word problems on the lesson practice pages, as they model the different kinds of language that may be used for the new concept just studied. For example, "altogether" indicates addition. These "key words" can be useful clues, but they should not be a substitute for understanding.

5. Look for practical applications that use the concept and ask questions in that context.

6. Have the student invent word problems to illustrate number problems from the lesson.

Cautions:

1. Unneeded information may be included in the problem. For example, we may be told that Suzie is eight years old, but the eight is irrelevant when adding up the number of gifts she received.

2. Some problems may require more than one step to solve. Model these questions carefully.

3. There may be more than one way to solve some problems. Experience will help the student choose the easier or preferred method.

4. Estimation is a valuable tool for checking an answer. If an answer is unreasonable, it is possible that the wrong method was used to solve the problem.

Place Value
with Expanded and Exponential Notation

In the base 10, or decimal, system, all numbers are represented by the digits 0 to 9 and by the place values units, tens, hundreds, thousands, etc. There are several ways to take a number and break it down into these components to make our normal system of *decimal notation* more explicit. We will explore three of them. The first way, which should be review, is *place-value notation*. In that notation, 156 is 100 + 50 + 6. The second way is *expanded notation*. In expanded notation, 156 is $1 \times 100 + 5 \times 10 + 6 \times 1$. It shows the individual digits multiplied by each place value. The third way is *exponential notation*. Using what we've learned about exponents, we can express all the place values in expanded notation as 10 to a power.

In figure 1 on the next page, we show several of the place values, beginning with the units place and continuing through the ten thousands place. The first three are easily shown with the blocks as in the figure, but the thousands place poses a challenge. One way to build 1,000 is a cube that is 10 by 10 by 10. (See figure 2.) If we follow the progression where the exponent indicates the dimension, using the second power for two dimensions and the third power for three dimensions, then we can't show 10,000. This is because 10,000 would be the fourth dimension, and we can't draw or construct a four-dimensional figure. Instead, we show each of the place values two-dimensionally. The number 1,000 is shown as 10 by 100. The number 10,000 is 100 by 100. If we had enough space, we could even show 100,000 as 100 by 1,000 and 1,000,000 as 1,000 by 1,000.

Figure 1

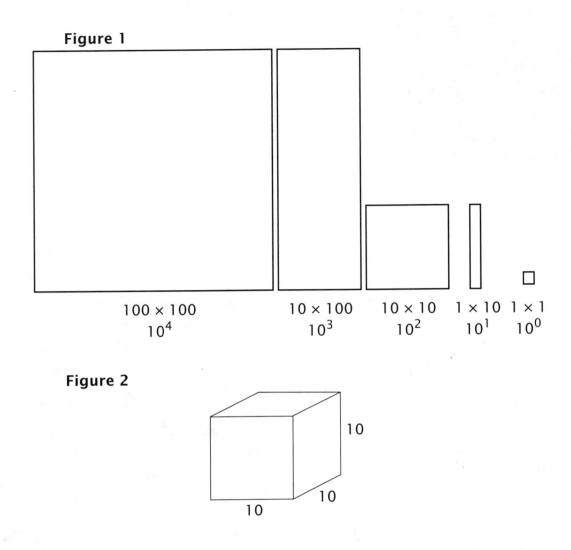

100×100	10×100	10×10	1×10	1×1
10^4	10^3	10^2	10^1	10^0

Figure 2

10
10
10

Notice how the place value corresponds to a power of 10. We will prove how $10^0 = 1$ in *Algebra 1*. Notice how the exponent for each power of 10 corresponds to the number of zeros in the place value. When you multiply by 10, another zero is added. 10^3 is 1,000, which has three zeros, and 10^1 is 10, which has one zero. Using this knowledge, we'll show a number first in expanded notation and then in exponential notation.

Figure 3

_____ , _____ _____ _____ , _____ ____ ____

1,000,000	100,000	10,000	1,000	100	10	1
10^6	10^5	10^4	10^3	10^2	10^1	10^0

Example 1

Express 245 in expanded notation and in exponential notation.

$$245 = 2 \times 100 + 4 \times 10 + 5 \times 1$$
$$245 = 2 \times 10^2 + 4 \times 10^1 + 5 \times 10^0$$

Example 2

Express 1,759 in expanded notation and in exponential notation.

$$1{,}759 = 1 \times 1{,}000 + 7 \times 100 + 5 \times 10 + 9 \times 1$$
$$1{,}759 = 1 \times 10^3 + 7 \times 10^2 + 5 \times 10^1 + 9 \times 10^0$$

Example 3

Express 803 in expanded notation and in exponential notation.

$$803 = 8 \times 100 + 3 \times 1$$
$$803 = 8 \times 10^2 + 3 \times 10^0$$

Example 4

Express $5 \times 1{,}000 + 6 \times 100 + 4 \times 10$ in decimal notation.

$$5 \times 1{,}000 + 6 \times 100 + 4 \times 10 = 5{,}640$$

Decimal Numbers with Expanded Notation

Decimal numbers are fractions written in the base 10 system. They are a blend of what we know of place value and of fractions. If you keep this in mind throughout our study of decimals, you will be in good shape. This lesson will go back and forth between fractions and place value to show how decimals relate to all we have learned up to this point. We are interested not only in learning how to add, subtract, multiply, and divide decimals, but also in learning the concepts that shape our understanding of this subject.

Some teachers refer to decimals as *decimal fractions.* This is because decimals are fractions written in the same base 10, or decimal notation, which we have already been using for whole numbers. In figure 1, a fraction is changed to an equivalent fraction with a denominator of 10. After constructing 2/5 with the overlays, place the clear overlay with 10 spaces to make tenths on top of 2/5. The result is 4/10 or four-tenths.

Figure 1

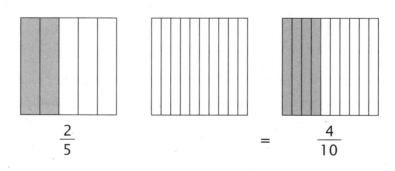

$$\frac{2}{5} \qquad = \qquad \frac{4}{10}$$

In expanded notation, 34 is $3\times10 + 4\times1$. The digits are 3 and 4, and the values are ten and one. A fraction may also be written in expanded notation. The fraction 2/5 can be written as $2\times\frac{1}{5}$. The digit 2 tells how many, and the value 1/5 tells what kind. In the decimal system, there is no fifths place. In the past, we have not dealt with any place value less than one (units). Now we need some smaller place values.

Recall that as you move from right to left to increase the place values, you multiply by a factor of 10. As you decrease and move from the larger to the smaller values and from left to right, you divide by a factor of 10. See figure 2.

Figure 2

100,000	10,000	1,000	100	10	1
$10\cdot10{,}000$	$10\cdot1{,}000$	$10\cdot100$	$10\cdot10$	$10\cdot1$	

100,000	10,000	1,000	100	10	1
	$100{,}000\div10$	$10{,}000\div10$	$1{,}000\div10$	$100\div10$	$10\div10$

Since we are looking for values less than one, or fractional values, we need to continue this pattern of decreasing by, or dividing by, a factor of 10. From studying fractions, we know that dividing by 10 is the same as multiplying by 1/10, since the symbol "/10" means "divided by 10." Picking up where we left off in figure 2, we'll divide 1 by 10 or multiply 1 by 1/10. See figure 3.

Figure 3

1	$\frac{1}{10}$	$\frac{1}{100}$	$\frac{1}{1{,}000}$
	$1\div10$	$(1/10)\div10$	$(1/100)\div10$
	$1\times(1/10)$	$(1/10)\times(1/10)$	$(1/100)\times(1/10)$

All we need now is a symbol to separate the whole-number place values from the fractional place values. As in money, the symbol is a decimal point. Between the units place and the tenths place, we put a decimal point. See figure 4 for our new fraction values written along with the numbers for units through hundreds. Notice that the units place, and not the decimal point, is the pivotal place.

Figure 4

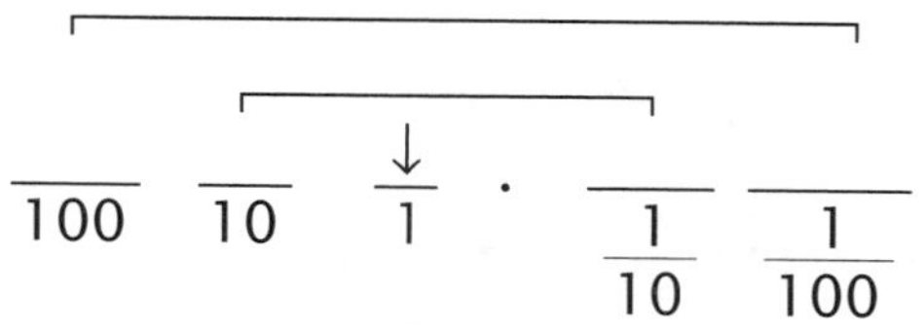

$$\overline{100}\quad\overline{10}\quad\overset{\downarrow}{\overline{1}}\;\cdot\;\overline{\dfrac{1}{10}}\quad\overline{\dfrac{1}{100}}$$

Since we have values that are less than one unit, refer back to figure 2, where we changed 2/5 to 4/10. We did this because there is not a 1/5 place in the decimal system. However, there is a 1/10 place. In expanded notation, 4/10 is written as $4\times\frac{1}{10}$. The fraction 4/10 may also be written as the decimal 0.4 (figure 5). It can be expressed as 4 in the tenths place, or 4 tenths, or 0.4. The zero added to the left of the decimal point shows us that there are no other values in the unit place. It calls attention to the unit place as well as the decimal point.

Figure 5

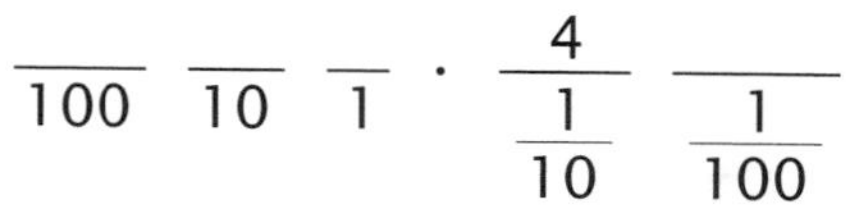

$$\overline{100}\quad\overline{10}\quad\overline{1}\;\cdot\;\overset{4}{\overline{\dfrac{1}{10}}}\quad\overline{\dfrac{1}{100}}$$

Money is one place where we commonly use decimal notation to show fractions. If one unit is one dollar, 1/10 of a dollar is one dime, and 1/100 of a dollar (or 1/10 of a dime) is one cent, or one penny. Two-fifths of a dollar is the same as 4/10 of a dollar or four dimes, as shown in figure 5. We will discuss more about money on the worksheets and in another lesson.

Study the examples below to add to your understanding of decimals.

Example 1
Express 34.762 with expanded notation.

$$34.762 = 3\times10 + 4\times1 + 7\times\dfrac{1}{10} + 6\times\dfrac{1}{100} + 2\times\dfrac{1}{1,000}$$

Example 2
Express 590.184 with expanded notation.

$$590.184 = 5\times100 + 9\times10 + 1\times\dfrac{1}{10} + 8\times\dfrac{1}{100} + 4\times\dfrac{1}{1,000}$$

Because of what we know of exponents, we can rewrite examples 1 and 2 in exponential notation. Remember that 10^0 is used to represent the units place. We'll learn why $10^0 = 1$ in Algebra 1.

Example 1   (in exponential notation)

$$34.762 = 3 \times 10^1 + 4 \times 10^0 + 7 \times \frac{1}{10^1} + 6 \times \frac{1}{10^2} + 2 \times \frac{1}{10^3}$$

Example 2   (in exponential notation)

$$590.184 = 5 \times 10^2 + 9 \times 10^1 + 1 \times \frac{1}{10^1} + 8 \times \frac{1}{10^2} + 4 \times \frac{1}{10^3}$$

Add Decimal Numbers

In this lesson, we will begin using the algebra-decimal inserts to represent decimals. Turn a red hundred square upside down so the hollow side is showing, and snap the flat green piece (from the algebra/decimal inserts) into the back. Then turn over several blue 10 bars and snap the flat blue pieces (also from the inserts) into their backs. Then take out the little one-half inch red cubes.

The large green square represents one unit. We've increased the size of the unit from the little green cube to this larger size, just as we did when learning fractions. Since the large green square represents one, what do you think the flat blue bars represent? It takes ten of them to make one, so they are each 1/10 or 0.1. The red cubes represent 1/100 or 0.01.

In figure 1, we show how to represent 1.56 or $1 \times 1 + 5 \times \frac{1}{10} + 6 \times \frac{1}{100}$ with the decimal inserts.

Figure 1

$$1 \quad . \quad 5 \quad 6 \quad = 1.56$$

As we've said before, decimal notation is used for money. If figure 1 represents money with the green unit as one dollar, then 1/10 of a dollar is one dime and is represented by the blue 1/10 bars. As shown with the red cubes, 1/100 of a dollar or 1/10 of a dime is one cent, or one penny.

Figure 2

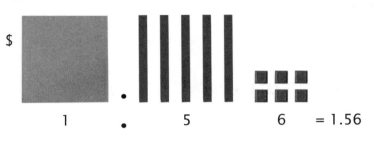

When you add up the change in your pocket, you add dollars to dollars, dimes to dimes, and cents to cents. Thinking of how we count money helps us to understand why the key concept for adding and subtracting decimals is that we add units to units, tenths to tenths, hundredths to hundredths, and thousandths to thousandths. This shouldn't seem strange because we have been adding like place values, regrouping as necessary, since *Beta*.

The easiest way to distinguish the values and make sure you are combining like values is by writing the problem vertically, so the decimal point in one number is directly above (or below) the decimal point in the other number. Lining up these points ensures you that your place values are also lined up. You may only add or subtract two numbers if they have the same value.

If zero is the only number to the left of the decimal point, you may either keep it or omit it when setting up a problem for computation.

When using the inserts, it is clear that you can only add the green to the green, the blue to the blue, etc. When you don't have enough inserts for larger numbers, always line up the decimal points. The same skills are used for adding decimals and money as for adding any number. Remember that decimals are base 10. You've just learned some new kinds of decimal values.

Example 1
Add 1.56 + 1.23

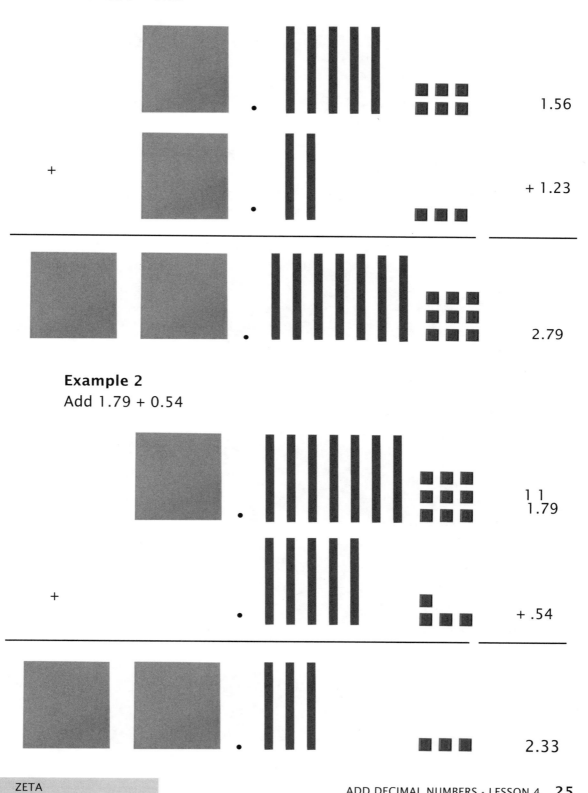

1.56

+   + 1.23

2.79

Example 2
Add 1.79 + 0.54

1 1
1.79

+   + .54

2.33

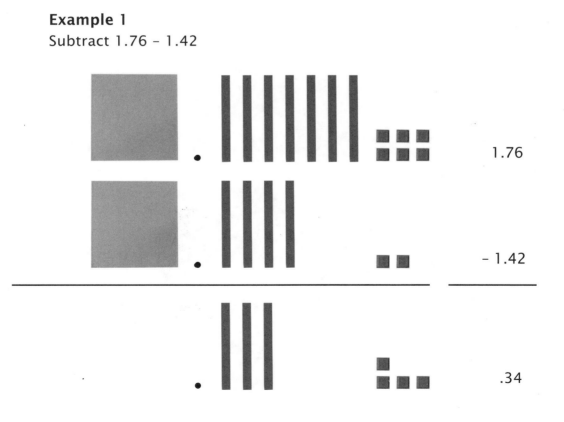

LESSON 5

Subtract Decimal Numbers

When subtracting decimals, the rules of regrouping still apply once you have lined up the place values. It helps to think of money when adding and subtracting decimals. In examples 1 and 2, subtraction problems are illustrated. There are no blocks or inserts to represent subtraction, but at this stage the student should be able to subtract accurately and readily.

Example 1
Subtract 1.76 – 1.42

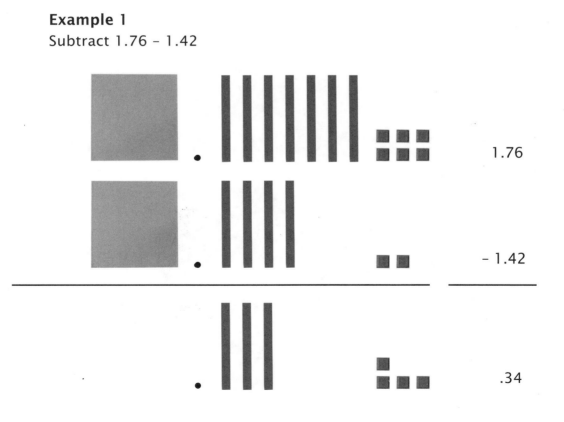

$$1.76$$
$$- 1.42$$
$$.34$$

Example 2
Subtract 2.29 – 1.75

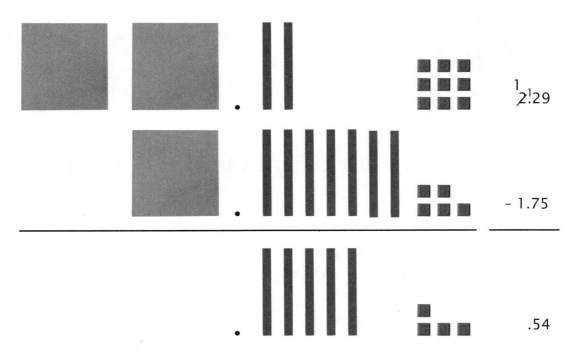

$$\begin{array}{r} \overset{1}{2}.\overset{1}{2}9 \\ -1.75 \\ \hline .54 \end{array}$$

Metric System Origin–Meter, Liter, Gram
and Greek Prefixes; Multi-step Word Problems

The term metric comes from the unit of length, the *meter*. ("Meter" comes from the word "metron" which refers to a device for measuring). In the 1790s a group of French scientists were asked to create a consistent system of measurement. They decided to start with the distance from the North Pole to the equator and to divide it into 10,000,000 equal pieces. Each of these pieces was defined as one "mètre" (in French). We write this as "meter" in American English. Today we have more accurate ways of measuring the earth, and we define a meter as the distance light travels in a certain fraction of a second.

The metric system also uses different base units to measure things other than distance. The *gram*, for example, is used to measure mass (which also determines weight), and the *liter* is used to measure volume and capacity.

Rather than having several different units that measure the same thing (like using gallons, bushels, and teaspoons to measure volume), the metric system uses prefixes which can be put in front of any base unit to describe something larger or smaller. The three prefixes used in this lesson come from the Greek words for "thousand", "hundred", and "ten," respectively. Putting "kilo" before any base unit creates a unit that is equal to 1,000 of the base unit. For example, a kilogram is equal to 1,000 grams, and a kiloliter is equal to 1,000 liters. Each of the prefixes works the same way, and each of them (like each base unit) has an abbreviation.

Prefixes and Abbreviations
    kilo (k) = 1,000
    hecto (h) = 100
    deka (da)= 10

Base Units and Abbreviations

meter (m)	indicates distance or length
liter (l or L)	indicates volume and capacity
gram (g)	indicates mass and weight

You can combine these abbreviations in the same way that you combine the prefixes and base units. For example, you can combine the "k" (for kilo) and "m" (for meter) to get km (for kilometer). 1 km = 1,000 m.

Be aware that in some places deka is spelled deca. Some older books use different abbreviations for deka like "dk" or "D", but "da" is the official abbreviation. You should not use a lowercase "d" by itself because that stands for the prefix "deci," which you will be learning in the next lesson. Deka is the only metric prefix that has a two-letter abbreviation.

Liter can be abbreviated either with an uppercase "L" or a lowercase "l." We usually use an uppercase "L" when the abbreviation is written by itself to avoid confusion with the numeral "1". You can use the lowercase "l" when it is combined with another abbreviation, such as "kl" for kiloliter.

As we learn these units we can write the relationship between each prefixed unit and its base unit as an equation, like 1 kilometer = 1,000 meters, or as a ratio:

$$\frac{1 \text{ kilometer}}{1,000 \text{ meters}}$$

You will learn more about ratios in enrichment pages 11G, 12G, 13G, and 14G. If you find it helpful, you could do some of these pages before returning to lesson 6. All of the ratios you will be writing in this lesson are equal to 1 because the amount on the top and the amount on the bottom are equal to each other. Understanding this will be important in lesson 8 when you start converting between metric units.

Example 1
Fill in the spaces above and below the lines with the appropriate word, number, and abbreviation.

1. $\dfrac{1 \text{ kilogram (kg)}}{1,000 \text{ grams (g)}}$ $\dfrac{\rule{3cm}{0.4pt}}{100 \text{ grams (g)}}$ $\dfrac{\rule{3cm}{0.4pt}}{10 \text{ grams (g)}}$ $\dfrac{\rule{3cm}{0.4pt}}{1 \text{ gram (g)}}$

2. _____ | _____ | _____ | _____
| | 100 liters (L) | | 1 liters (L)

3. _____ | _____ | 1 dekameter (dam) | 1 meter (m)
1,000 meter (m) | | |

Example 1 solution

1.	1 kilogram (kg)	1 hectogram (hg)	1 dekagram (dag)	1 gram (g)
	1,000 grams (g)	100 grams (g)	10 grams (g)	1 gram (g)
2.	1 kiloliter (kl)	1 hectoliter (hl)	1 dekaliter (dal)	1 liter (L)
	1,000 liters (L)	100 liters (L)	10 liters (L)	1 liter (L)
3.	1 kilometer (km)	1 hectometer (hm)	1 dekameter (dam)	1 meter (m)
	1,000 meter (m)	100 meter (m)	10 meter (m)	1 meter (m)

Multi-step Word Problems

The student workbook includes some fairly simple two-step word problems. Some students may be ready for more challenging problems. Here are a few to try, along with some tips for solving this kind of problem. You may want to read and discuss these with your student as you work out the solutions together. The purpose is to stretch, not to frustrate. If you do not think the student is ready, you may want to come back to these later.

There are more multi-step word problems on the next page and in lessons 12, 18, and 24 of this instruction manual. The answers are at the end of the solutions at the back of this book.

1. Jill bought items costing $3.45, $1.99, $6.59, and $12.98. She used a coupon worth $2.50. If Jill had $50.00 when she went into the store, how much money did she have when she left?

Although the problem asks only one question, there are other questions that must be answered first. The key to solving the problem is determining what the unstated questions are. Since the final question is asking for the leftover money, the unstated questions are: "What is the total of Jill's purchases?" and "What was the total bill after using the coupon?"

You might make a list of questions like this:

1. Total of purchases?

2. Total bill after subtracting coupon?

3. Leftover money?

2. Luke and Seth started out to visit Uncle Arnie. After driving 50 miles, they saw a restaurant, and Luke wanted to stop for lunch. Seth wanted to look for something better, so they drove on for eight miles before giving up and going back to the restaurant. After eating, they traveled on for 26 more miles from the restaurant. Seth saw a sign for a classic car museum, which they decided to visit. The museum was six miles from their route. After returning to the main road, they drove for another 40 miles and arrived at Uncle Arnie's house. How many miles is it from Luke and Seth's house to Uncle Arnie's house? How many miles did they drive on the way there?

   The key to solving this is a careful drawing. It does not have to be to scale, but it should include all the parts of the journey.

3. Last week we got 3.5 inches of snow. Six-tenths of an inch melted before another storm added 8.3 inches. Since then we have lost 4.2 inches to melting or evaporation. How many inches of snow are left on the ground?

Metric System–Latin Prefixes

In the last lesson you learned three prefixes that come from Greek and are used to create larger units of measure. In this lesson you will learn three prefixes which come from Latin and are used to create smaller units. *Milli* means one thousandth, while *centi* and *deci* mean one hundredth and one tenth. You can abbreviate them with the letters m, c, and d.

$$\text{deci (d)} = \frac{1}{10}$$

$$\text{centi (c)} = \frac{1}{100}$$

$$\text{milli (m)} = \frac{1}{1,000}$$

One way to avoid confusing these with kilo, hecto, and deka is to notice that deci and centi have a soft "c" sound like the "s" in "small." Another fact which may help is that American money uses a modified version of two of these prefixes. A decidollar (one tenth of a dollar) is called a dime ("dime" is the French form of the Latin decima) and a centidollar or hundredth of a dollar is called a cent. If you think "dime" when you see "deci" it may help you avoid confusing it with "deka."

Just like the prefixes from lesson 6, these prefixes can be combined with the base units. You can also combine the abbreviations. 100 centimeters (cm) = 1 meter (m). Remember that a liter by itself is abbreviated with a capital "L" but when combined with a prefix, it may be represented by either an uppercase or a lowercase "l". Milliliters, for example, can be written as mL or as ml.

Example 1

Fill in the spaces above and below the lines with the appropriate word, number and abbreviation.

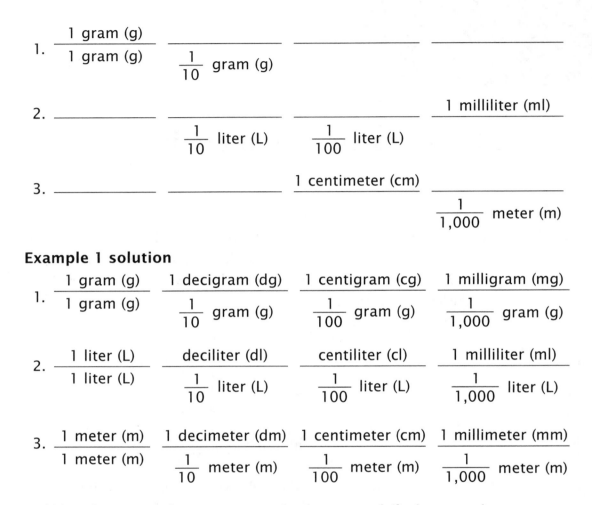

1. $\dfrac{1 \text{ gram (g)}}{1 \text{ gram (g)}}$ $\dfrac{\rule{2cm}{0.4pt}}{\frac{1}{10} \text{ gram (g)}}$ $\dfrac{\rule{2cm}{0.4pt}}{\rule{2cm}{0.4pt}}$ $\dfrac{\rule{2cm}{0.4pt}}{\rule{2cm}{0.4pt}}$

2. $\dfrac{\rule{2cm}{0.4pt}}{\rule{2cm}{0.4pt}}$ $\dfrac{\rule{2cm}{0.4pt}}{\frac{1}{10} \text{ liter (L)}}$ $\dfrac{\rule{2cm}{0.4pt}}{\frac{1}{100} \text{ liter (L)}}$ $\dfrac{1 \text{ milliliter (ml)}}{\rule{2cm}{0.4pt}}$

3. $\dfrac{\rule{2cm}{0.4pt}}{\rule{2cm}{0.4pt}}$ $\dfrac{\rule{2cm}{0.4pt}}{\rule{2cm}{0.4pt}}$ $\dfrac{1 \text{ centimeter (cm)}}{\rule{2cm}{0.4pt}}$ $\dfrac{\rule{2cm}{0.4pt}}{\frac{1}{1,000} \text{ meter (m)}}$

Example 1 solution

1. $\dfrac{1 \text{ gram (g)}}{1 \text{ gram (g)}}$ $\dfrac{1 \text{ decigram (dg)}}{\frac{1}{10} \text{ gram (g)}}$ $\dfrac{1 \text{ centigram (cg)}}{\frac{1}{100} \text{ gram (g)}}$ $\dfrac{1 \text{ milligram (mg)}}{\frac{1}{1,000} \text{ gram (g)}}$

2. $\dfrac{1 \text{ liter (L)}}{1 \text{ liter (L)}}$ $\dfrac{\text{deciliter (dl)}}{\frac{1}{10} \text{ liter (L)}}$ $\dfrac{\text{centiliter (cl)}}{\frac{1}{100} \text{ liter (L)}}$ $\dfrac{1 \text{ milliliter (ml)}}{\frac{1}{1,000} \text{ liter (L)}}$

3. $\dfrac{1 \text{ meter (m)}}{1 \text{ meter (m)}}$ $\dfrac{1 \text{ decimeter (dm)}}{\frac{1}{10} \text{ meter (m)}}$ $\dfrac{1 \text{ centimeter (cm)}}{\frac{1}{100} \text{ meter (m)}}$ $\dfrac{1 \text{ millimeter (mm)}}{\frac{1}{1,000} \text{ meter (m)}}$

Including fractions in these ratios can make these ratios difficult to use when solving problems. Here is a modified version which we will be using in lesson 8.

$\dfrac{1 \textbf{ kilo}\text{unit}}{1,000 \text{ units}}$ $\dfrac{1 \textbf{ hecto}\text{unit}}{100 \text{ units}}$ $\dfrac{1 \textbf{ deka}\text{unit}}{10 \text{ units}}$ $\dfrac{1 \text{ unit}}{1 \text{ unit}}$ $\dfrac{10 \textbf{ deci}\text{units}}{1 \text{ unit}}$ $\dfrac{100 \textbf{ centi}\text{units}}{1 \text{ unit}}$ $\dfrac{1,000 \textbf{ milli}\text{units}}{1 \text{ unit}}$

Comparing Metric Units

One liter is exactly one cubic decimeter or 1,000 cubic centimeters.

$$1 \text{ L} = 1 \text{ dm}^3 = 1,000 \text{ cm}^3$$

One cubic meter is exactly one kiloliter or 1,000 liters.

$$1 \text{ m}^3 = 1 \text{ kl} = 1,000 \text{ L}$$

One milliliter is exactly one cubic centimeter.

$$1 \text{ ml} = 1 \text{ cm}^3$$

One liter of water weighs approximately one kilogram or 1,000 grams.

One milliliter, or one cubic centimeter, of water weighs about one gram.

It may also help to note that the inside of a green Math-U-See unit block is close to one cubic centimeter or one milliliter, so one gram of water would come close to filling up one green unit block.

Comparing with Customary Units

You may wonder how some of the common metric units correspond to familiar U. S. customary units. Here are a few that I think are worth remembering.

One meter is a little over three feet (ft) or one yard (yd).

One meter is about 39.37 inches (in); one yard is 36 inches.

One centimeter is less than 1/2, or about 0.4 of an inch.

One inch is exactly 2.54 cm, or approximately 2½ cm.

One liter is approximately 1.06 quarts (qt).

One ounce (oz) is approximately 28 grams.

One gram is about the weight of a small paper clip.

One kilogram is approximately 2.2 pounds (lb).

One kilometer is a little over 1/2 mile (mi), or approximately 0.6 of a mile.

LESSON 8

Metric System Conversion–Part 1

In the last two lessons you learned the meaning of these metric prefixes, as well as their abbreviations.

kilo (k) = 1,000

hecto (h) = 100

deka (da) = 10

deci (d) = $\dfrac{1}{10}$

centi (c) = $\dfrac{1}{100}$

milli (m) = $\dfrac{1}{1,000}$

You have also been writing these relationships as ratios in tables like these:

Figure 1

| $\dfrac{1\ \textbf{kilo}\text{unit}}{1,000\ \text{units}}$ | $\dfrac{1\ \textbf{hecto}\text{unit}}{100\ \text{units}}$ | $\dfrac{1\ \textbf{deka}\text{unit}}{10\ \text{units}}$ | $\dfrac{1\ \text{unit}}{1\ \text{unit}}$ | $\dfrac{1\ \textbf{deci}\text{units}}{\frac{1}{10}\ \text{unit}}$ | $\dfrac{1\ \textbf{centi}\text{units}}{\frac{1}{100}\ \text{unit}}$ | $\dfrac{1\ \textbf{milli}\text{units}}{\frac{1}{1,000}\ \text{unit}}$ |

| $\dfrac{1\ \textbf{kilo}\text{unit}}{1,000\ \text{units}}$ | $\dfrac{1\ \textbf{hecto}\text{unit}}{100\ \text{units}}$ | $\dfrac{1\ \textbf{deka}\text{unit}}{10\ \text{units}}$ | $\dfrac{1\ \text{unit}}{1\ \text{unit}}$ | $\dfrac{10\ \textbf{deci}\text{units}}{1\ \text{unit}}$ | $\dfrac{100\ \textbf{centi}\text{units}}{1\ \text{unit}}$ | $\dfrac{1,000\ \textbf{milli}\text{units}}{1\ \text{unit}}$ |

In this lesson you will use these ratios as multiplication factors to convert larger units to smaller units. You will first learn the standard method, which involves multiplying by conversion ratios, and then we will look at an alternative method and a shortcut.

Method 1 (Standard Method)

The standard way of converting units using a ratio is to multiply the original unit by the appropriate ratio. For example if you need to convert one kilogram to grams, you multiply by the ratio which shows the relationship between kilograms and grams. Because the amounts on the top and bottom of these ratios are the same, you can present each ratio either way without changing the value.

Figure 2

$$\frac{1 \text{ kilogram (kg)}}{1,000 \text{ grams (g)}} = \frac{1,000 \text{ grams (g)}}{1 \text{ kilogram (kg)}}$$

When choosing which ratio to multiply by, you will need to arrange it so that the unit you are converting to is on top.

Example 1

$$1 \text{ kilogram (kg)} \times \frac{1,000 \text{ m}}{1 \text{ kilogram (kg)}} = 1,000 \text{ grams (g)}$$

Notice that you can cancel the unit kilograms because we are dividing kilograms by kilograms, and because of this you will end up with an answer in grams. Also notice that when converting kilograms to grams you ended up multiplying by 1,000 which is exactly what "kilo" means.

It becomes slightly more complicated when you convert from one unit that has a prefix to another unit which has a prefix. For example, if you convert 1 kilometer to centimeters it requires two of the conversion ratios as factors.

Example 2

$$1 \text{ km} \times \frac{1,000 \text{ m}}{1 \text{ km}} \times \frac{100 \text{ cm}}{1 \text{ m}} = 1(1,000)(100 \text{ cm}) = 1,000,000 \text{ cm}$$

If you start with a number of units other than one, you just multiply that number by each factor along the way.

Example 3

$$15 \, \cancel{hl} \times \frac{100 \, \cancel{l}}{1 \, \cancel{hl}} \times \frac{10 \, dl}{1 \, \cancel{l}} = 15(100)(10)dl = 15,000 \, dl$$

If you need to convert between two units that are close to each other in size, you may end up both multiplying and dividing. For example if you use the ratios to convert one centiliter to milliliters it looks like this.

Example 4

$$1 \, \cancel{cl} \times \frac{1 \, \cancel{l}}{10 \, \cancel{cl}} \times \frac{1,000 \, ml}{1 \, \cancel{l}} = 10 \, ml$$

Method 2 (Alternative Method)

Think about the way conversions like this interact with place value. When you go from tens to hundreds you multiply by 10, and when you go from hundreds to thousands you multiply by 10 again. Because of this, if you create a ratio between any two units that are next to each other on our original table, you can describe them as a ratio of 10 to 1.

Figure 3

$\dfrac{10 \text{ hecto}\text{units}}{1 \text{ kilo}\text{unit}}$	$\dfrac{10 \text{ deka}\text{units}}{1 \text{ hecto}\text{unit}}$	$\dfrac{10 \text{ units}}{1 \text{ deka}\text{unit}}$	$\dfrac{10 \text{ deci}\text{units}}{1 \text{ unit}}$	$\dfrac{10 \text{ centi}\text{units}}{1 \text{ deci}\text{unit}}$	$\dfrac{10 \text{ milli}\text{units}}{1 \text{ centi}\text{unit}}$

If you are converting between two units that are adjacent to each other, these 10 to 1 ratios can be very helpful.

Example 5

$$3 \, \cancel{km} \times \frac{10 \, hm}{1 \, \cancel{km}} = 30 \, hm$$

$$2 \, \cancel{cm} \times \frac{10 \, mm}{1 \, \cancel{cm}} = 20 \, mm$$

If you are converting between units that are farther apart, you end up multiplying by the same ratio several times. You use the ratio once for each step that the units were separated by on the original chart. As shown in the video lesson, this is the same as multiplying by 10 for each step from larger to smaller on the chart.

Example 6

$$27 \ \cancel{hg} \ \times \ \frac{10 \ \cancel{dag}}{1 \ \cancel{hg}} \times \frac{10 \ \cancel{g}}{1 \ \cancel{dag}} \times \frac{10 \ dg}{1 \ \cancel{g}} \ = 27(10)(10)(10) \ dg = 27{,}000 \ dg$$

Method 2 Shortcut

Each place in our decimal place value system is 10 times the place before it. Each metric unit is also 10 times as large as the unit to the right of it on the chart. Because of this, we are, in effect, adding a zero for each step along the unit conversion chart as we go from larger to smaller.

In summary, you have now learned two different ways for converting a larger metric unit to a smaller metric unit.

Example 7 (Method 1)

$$32 \ \cancel{hl} \ \times \ \frac{100 \ \cancel{l}}{1 \ \cancel{hl}} \times \frac{10 \ dl}{1 \ \cancel{l}} \ = 32(100)(10) \ dl = 32{,}000 \ dl$$

Example 8 (Method 2)

$$32 \ \cancel{hg} \ \times \ \frac{10 \ \cancel{dag}}{1 \ \cancel{hg}} \times \frac{10 \ \cancel{g}}{1 \ \cancel{dag}} \times \frac{10 \ dg}{1 \ \cancel{g}} \ = 32(10)(10)(10) \ dg = 32{,}000 \ dg$$

Because each step in the alternative method involves multiplying by 10 as you go from larger to smaller, you can use the shortcut of just changing the place value of the original number, thus "moving" the decimal place by adding a zero for each conversion step.

Example 9 (Method 2 Shortcut)

32. → 32000.

When doing the problems in the worksheets, you may use whichever method works best for you in a given problem.

LESSON 9

Multiply by 1/10 or 0.1

Multiplying decimals is based on what we already know of double-digit multiplication. Only our dimensions have changed. Recognizing the dimensions and area of the inserts is important. Understanding the difference between the factors and the product is also critical. If needed, review the introduction to the decimal inserts in lesson 4.

The dimensions of the green unit are one over by one up, so the area is one. It represents a problem with the factors one by one and a product of one. The flat blue bar represents a product of 1/10 or 0.1, with an over factor of 1/10 and an up factor of one. Look at the figures below and think about this until you feel comfortable with the concept. Once you understand the blue bar, the red block will make more sense. It represents a product of 1/100 or 0.01, and the factors are 1/10 by 1/10, or 0.1×0.1. There is more explanation on the next page.

Figure 1

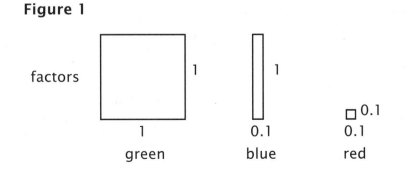

Figure 2

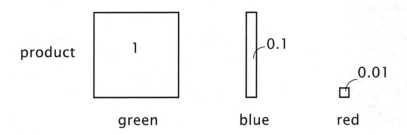

These figures look simple, but they are the key to understanding how to multiply decimals. Multiply the factors shown in figure 1 to find the product in figure 2. The unit block's factors are one and one, so one times one equals one. For the blue 1/10 bar, the factors are 1/10 and one, and 1/10 times one equals 1/10. With decimals this is $0.1 \times 1 = 0.1$.

The red 1/100 piece has the factors 1/10 and 1/10. Using fractions, you are taking 1/10 of 1/10. In the three-step process for taking a fraction of a fraction, you divide 1/10 into ten equal parts, and then take one of them. The result of 1/10 times 1/10 is 1/100. With decimals, it is $0.1 \times 0.1 = 0.01$. Money can help here. We know that 1/10 of a dime (which is 1/10 of a dollar) is a cent.

Example 1
Multiply 1.2 times 0.4

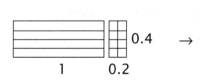 $\rightarrow$

Place Value Notation	Regular Notation
$1 + .2$	1.2
$\times\ .4$	$\times\ .4$
$.4\ +.08$	$.48$

Example 2
Multiply 2.3 times 0.2

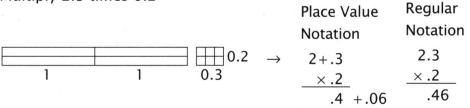 $\rightarrow$

Place Value Notation	Regular Notation
$2 + .3$	2.3
$\times\ .2$	$\times\ .2$
$.4\ +.06$	$.46$

Example 3

Find the factors and area of this rectangle.

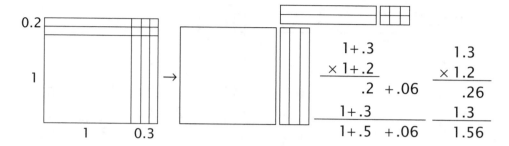

$$\begin{array}{r} 1+.3 \\ \times\ 1+.2 \\ \hline .2\ \ +.06 \\ 1+.3 \\ \hline 1+.5\ \ +.06 \end{array} \qquad \begin{array}{r} 1.3 \\ \times\ 1.2 \\ \hline .26 \\ 1.3 \\ \hline 1.56 \end{array}$$

The factors are 1.3 by 1.2. The area is 1.56.

Notice that when writing the problem, the up factor is on the line with the multiplication symbol, and the over factor is on the top line. Switching these factors will still produce the same answer, but it won't correspond with the picture shown above.

I write this differently than most, so that the conceptual teaching comes first, and the rules follow the concept. I do this for two reasons.

First, most students expect a large answer when multiplying, as in $10 \times 20 = 200$. They usually see a product much larger than either factor. But don't forget, we are multiplying mixed numbers now, and our product is not going to be significantly larger than either factor.

Second, I want to show the reason for the rule normally employed when multiplying decimals, which is to multiply and then count the number of places to the right of the decimal point. Instead of moving the decimal point, we're going to do something "radical" and leave it stationary. Actually, moving decimal points is radical since they don't really move; it just appears that they do.

In the first step, 0.2×0.3 (or $\frac{2}{10} \times \frac{3}{10}$) shows that 0.2×0.3 is 0.06 (or 6/100). Placing the 6 in the hundredths place shows two things. First, our product will be small, since we are starting two places to the right, and second, this is where the "multiply and then count" comes from. Tenths (one place over) times tenths (one place over) yields hundredths in our answer. One place over plus one place over equals two places over.

Then 0.2×1 is 0.2, and 1×0.3 is 0.3, and 1×1 is 1. When adding, always add the same place values, so $0.06 + 0$ is 0.06, $0.2 + 0.3$ is 0.5, and $1 + 0$ is 1. The answer is 1.56, as predicted.

Example 4
Find the factors and area of this rectangle.

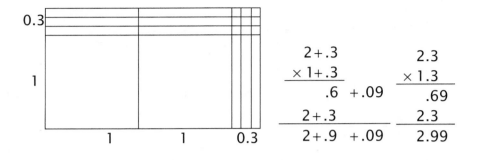

$$
\begin{array}{r}
2+.3 \\
\times\ 1+.3 \\
\hline
.6\ \ +.09 \\
2+.3 \\
\hline
2+.9\ \ +.09
\end{array}
\qquad
\begin{array}{r}
2.3 \\
\times\ 1.3 \\
\hline
.69 \\
2.3 \\
\hline
2.99
\end{array}
$$

Multiply Decimals by 1/100 or 0.01

Having mastered multiplying by tenths in the previous lesson, we are going to proceed to multiplying by hundredths. Emphasize thinking through each step and not merely memorizing. Even if the student chooses to multiply and then count as described later in the lesson, the thinking taught here is still invaluable for estimating the final answer. We'll begin by taking 1/100 or 0.01 of several numbers.

Example 1
Find 1/100 of 1. Remember that "of" means "times."

$$\frac{1}{100} \text{ of } 1 \text{ or } \frac{1}{100} \times \frac{1}{1} \text{ is } \frac{1}{100} \qquad 0.01 \times 1 = 0.01$$

Example 2
Find 1/100 of 100.

$$\frac{1}{100} \text{ of } 100 \text{ or } \frac{1}{100} \times \frac{100}{1} \text{ is } \frac{100}{100} \text{ or } 1 \qquad 0.01 \times 100 = 1$$

Example 3
Find 1/100 of 10.

$$\frac{1}{100} \text{ of } 10 \text{ or } \frac{1}{100} \times \frac{10}{1} \text{ is } \frac{10}{100} \text{ or } \frac{1}{10} \qquad 0.01 \times 10 = 0.1$$

Example 4
Find 1/100 of 1/10.

$$\frac{1}{100} \text{ of } \frac{1}{10} \text{ or } \frac{1}{100} \times \frac{1}{10} \text{ is } \frac{1}{1,000} \qquad 0.01 \times 0.1 = 0.001$$

Notice that when you multiply by 1/100, it is the same as dividing by 100. When you divide by 100, your answer moves two places to the right because it is that much smaller. It appears that the decimal point "moves" two places to the left. These are two ways of describing the same phenomenon. Multiplying by 1/10 (which is 0.1 in decimal form) or dividing by 10 moves the answer one place to the right, and the decimal point appears to move one place to the left.

Example 5

```
  2.3          23      one place
× .32        × 32    + two places
 .046   or    46
 .69          69
 .736        736

             .736     three places from the right
```

We know how to multiply; the key is place value. You can think, "hundredths times tenths is thousandths," and begin in the thousandths place with the partial products. Or, you can multiply and solve the problem as if there were no decimals. Then count how many spaces to the right of the decimal point in both factors (one space plus two spaces). Count from the right and put the decimal point after the third space from the right.

This is the same as thinking "hundredths times tenths is thousandths," multiplying as if there were no decimal points, and then placing the decimal point in the answer so the last digit is in the thousandths place.

Example 6

```
 300.          300    zero places
× .02         ×  2  + two places
 6.00   or    600
```

6.00   two places from the right

A hundredth times a hundred is a unit.

Example 7

```
 47.           47    zero places
× .33         × 33  + two places
 1.41   or    141
14.1          141
15.51        1551
```

15.51   two places from the right

A hundredth times a unit is a hundredth.

Example 8

```
 4.32          432    two places
× .65         ×  65 + two places
 .2160   or   2160
2.592         2592
2.8080       28080
```

2.8080   four places from the right

A hundredth times a hundredth is a ten-thousandth.

Finding a Percentage

In a previous lesson we changed a fraction to a decimal by placing the tenth overlay on top of it. We can go a step further and place the other tenth overlay on top of the first one at a 90° angle. See figure 1.

Here we change 2/5 to 4/10 to 40/100. Notice that $\frac{4}{10} = 0.4$ and $\frac{40}{100} = 0.40$.

Figure 1

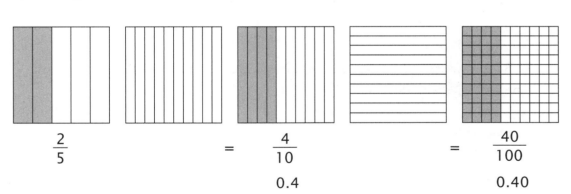

$$\frac{2}{5} \qquad = \qquad \frac{4}{10} \qquad = \qquad \frac{40}{100}$$

$$0.4 \qquad\qquad\qquad 0.40$$

In figure 2, we took the 40/100 and transformed it to a percentage by taking the one and the two zeros from the number 100 and changing them into a percent sign. We have changed a fraction (2/5) to a decimal (0.4) to a percentage (40%).

Figure 2

$$\frac{40}{100} \longrightarrow \frac{40/}{00} \longrightarrow \frac{40\%}{0} \longrightarrow 40\% \longrightarrow 40\%$$

Percent means "per hundred." A percentage is another way of writing hundredths. If you can change a fraction to hundredths (using either fraction or decimal notation), then you can easily change it to a percentage. The opposite is also true. If you have a percentage, you can immediately restate it as hundredths. To take a percentage of a number, simply change the percentage to hundredths and multiply it by the number. See example 1.

Example 1
Find 25% of 36.

25% = 0.25 and $0.25 \times 36 = 9$. So 25% of 36 is 9.

Another way to solve the same problem would be to change 25% to a fraction, and then reduce it and multiply to find a fraction of a number. See example 2.

Example 2
Find 25% of 36.

$$25\% = \frac{25}{100} = \frac{1}{4} \qquad \frac{1}{4} \times 36 = 9$$

Both ways are legitimate. Some problems are easier to solve using fractions and some are easier with decimals.

Example 3
Find the tax on an order of onion rings that cost $1.50 if the tax rate is 8%.

8% = 0.08 and $0.08 \times \$1.50 = \$0.12$

Example 4
Find the tip on a bill of $12.50 if you are tipping 16%.

16% = 0.16 and $0.16 \times \$12.50 = \$2.00$

Example 5

Burgers are 20% off on Tuesdays. How much is a $2.80 burger on Tuesday?

$20\% = 0.20$ and $0.20 \times \$2.80 = \$0.56$
$\$2.80 - \$0.56 = \$2.24$

Not to confuse the issue, but there is another way to solve example 5. If you are going to be taking 20% off, the final answer will be 100% minus 20% or 80% of the original amount. The total is 100%, or one. Eighty percent is 0.80 and $0.80 \times \$2.80 = \$2.24$. Instead of finding 20%, multiplying, and then subtracting, you are subtracting from 100% and then multiplying. Experiment by trying the same problem different ways to see which method is more comfortable for you.

There are several percentages that I have memorized that have proven helpful to me. I have listed them below in figure 3. We will be able to figure out some of the equivalents ourselves when we learn to divide decimals later in this book. Since one-fifth is 20%, then it follows that two-fifths is double that, or 40%, three-fifths is 60%, and four-fifths is 80%.

Figure 3

$$\frac{1}{1} = 1.00 = 100\%$$

$$\frac{1}{2} = 0.50 = 50\%$$

$$\frac{1}{4} = 0.25 = 25\%$$

$$\frac{3}{4} = 0.75 = 75\%$$

$$\frac{1}{5} = 0.20 = 20\%$$

Finding a Percentage > 100%
Word Problems

It is possible to have percentages greater than 100%. We know 100% is the same as one. Consider the percentage given in figure 1. Percentages greater than 100% are the same as improper fractions or other numbers larger than one.

Figure 1

$$1 = \frac{100}{100} = 100\% \quad 2 = \frac{200}{100} = 200\% \quad 3 = \frac{300}{100} = 300\%$$

Figure 2 shows percentages greater than one that can be written as mixed numbers rather than whole numbers.

Figure 2

$$1\frac{1}{2} = \frac{3}{2} = \frac{150}{100} = 150\% \text{ or}$$

$$1\frac{1}{2} = 1 + \frac{1}{2} = \frac{100}{100} + \frac{50}{100} = \frac{150}{100} = 150\%$$

$$2\frac{1}{4} = \frac{9}{4} = \frac{225}{100} = 225\% \text{ or}$$

$$2\frac{1}{4} = 2 + \frac{1}{4} = \frac{200}{100} + \frac{25}{100} = \frac{225}{100} = 225\%$$

In a problem that involves finding a tax, you can use the shortcut mentioned at the end of lesson 11 to save yourself some work. If onion rings are $1.50 with a tax rate of 8%, then the tax is 8% or $0.08 \times \$1.50$, which is $0.12. Adding $1.50 + $0.12 produces the actual cost of buying the onion rings, which is $1.62.

Instead of multiplying by 8%, and then adding, you could have multiplied by 1.08. You have to pay the total cost of the onion rings, which is 100% of the cost, plus the tax, which is 8% of the cost. So 108% is the cost of the rings and the tax inclusive. 100% is the same as 1.00 and 8% is the same as 0.08, so 108% is the same as 1.08. Multiply $1.50 times 1.08, and you get $1.62. Either method will work.

Example 1
Find the total cost for a meal priced at $12.50 if you are tipping 16% and the tax is 8%.

Total + tip + tax = 100% + 16% + 8% = 124%
$124\% \times \$12.50 = 1.24 \times \$12.50 = \$15.50$

Word Problems

Here are a few more multi-step word problems to try. You may want to read and discuss these with your student as you work out the solutions together. Again, the purpose is to stretch, not to frustrate. If you do not think the student is ready, you may want to come back to these later.

The answers are at the end of the solutions at the back of this book.

1. Naomi was ordering yarn from a website. If she ordered more than $50 worth of yarn labeled "discountable," she could take an extra 10% from the price of that yarn. Only some of the yarn she ordered was eligible for the discount. Here is what Naomi ordered:

   Two skeins at $2.50 per skein.   (not discountable)

   Eight skeins at $8.40 per skein.   (discountable)

   Seven skeins at $5.99 per skein.   (discountable)

   Tax is 5% and shipping is 8%. How much did Naomi pay for her yarn?

2. After Naomi received her order, she found that she could make a scarf from one skein of the yarn that cost $5.99 a skein. After taking into account the discount, taxes, and shipping for the yarn, how much would it cost to make five scarves?

3. Naomi made a scarf for her brother using the yarn that cost $2.50 per skein. He offered to pay Naomi for the yarn she used. If she used 1.5 skeins of yarn to make the scarf, what was the total cost of the yarn?

Reading Percentages in a Pie Graph

Circular *pie graphs* are often used to show percents. The whole pie is like the whole number one. If the whole pie is shaded, it represents 100%. If half of the pie is shaded, it represents one-half of 100% or 50%. Each shaded part of the pie illustrates a fraction or a percentage of the whole. In the following examples, you will see data displayed visually with a pie graph.

Example 1

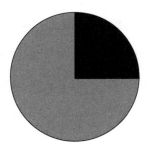

 25% of the book was illustrated with colored pictures.

75% of the book was black print on a white background.

Example 2

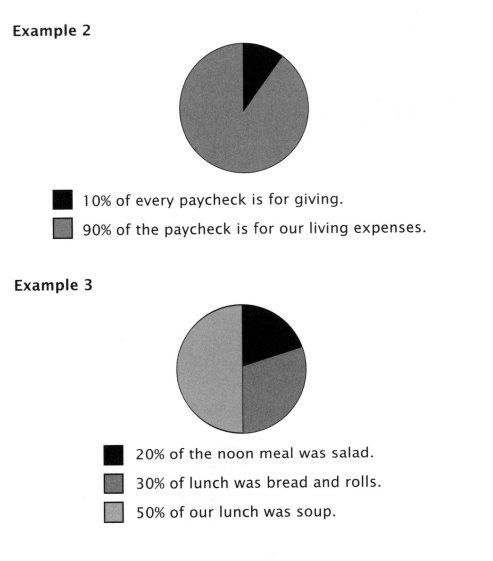

■ 10% of every paycheck is for giving.

▨ 90% of the paycheck is for our living expenses.

Example 3

■ 20% of the noon meal was salad.

▨ 30% of lunch was bread and rolls.

▨ 50% of our lunch was soup.

Example 4

■ 10% of the population are left-handed.

▨ 85% of the people are right-handed.

▨ 5% are ambidextrous.

Multiply All Decimals

What do you think will happen when you multiply by 1/1,000 or 0.001? The answer will move three places to the right, or the decimal point will appear to move three places to the left. This is because you are taking 1/1,000 of the original number. In the examples below, we'll estimate first by thinking what the answer should be and then work the problem using both methods. The first method is to line up the decimal points as in an addition problem to figure out where to begin, and to start your process from there. In example 1, two-hundredths times nine-hundredths is 18 ten-thousandths, so we begin four spaces to the right.

The other method is to count the spaces to the right of the decimal point in both factors, add them for the answer, and then multiply the numbers, paying attention only to the relative place value. In the answer, count the spaces from the right to place the decimal point. Either method will work.

Example 1

$$
\begin{array}{rl}
75.89 & \quad 7589 \quad \text{two place} \\
\times\ .52 & \quad \underline{\times\ 52} \ + \text{two places} \\
\hline
1.5178 \ \text{or} & \quad 15178 \\
\underline{37.945} & \quad \underline{37945} \\
39.4628 & \quad 394628 \\
& \quad 39.4628 \ \text{four places from the right}
\end{array}
$$

To estimate, round 75.89 to 80 and 0.52 to 0.5. Estimate: $80 \times 0.5 = 40$. Or, remember that 0.5 is the same as five-tenths, which is one-half, and one-half of 80 is 40. Either way, the answer should be approximately 40.

Example 2

```
    536.042              536042   three places
×      .08          ×         08 + two places
   42.88336  or     4288336
                    42.88336   five places from the right
```

To estimate, round 536.042 to 500 and 0.08 to 0.1. Since 500 times 0.1 is 50, the answer should be approximately 50.

Example 3

```
    703.091                703091   three places
×     1.584          ×        1584 + three places
    2.812364  or     2812364
   56.24728           5624728
  351.5455            3515455
  703.091             703091
1,113.696144        1113696144
                    1,113.696144  six places from the right
```

To estimate, round 703.091 to 700 and 1.584 to 2. Taking 700 times 2 gives 1,400. Or, think 1.584 is close to 1.5 or 1½, and 1½ times 700 is the same as 700 plus 350 or 1,050. Since 1.584 is between 1.5 and 2, the answer should be between our estimates of 1,050 and 1,400.

Metric System Conversions – Part 2

In Lesson 8, we learned two different methods for converting a larger unit to a smaller unit.

Example 1 (Standard Method)
Convert 32 hectoliters = _____ deciliters.

$$32 \; h\!l \; \times \; \frac{100 \; l}{1 \; h\!l} \; \times \; \frac{10 \; dl}{1 \; l} = 32(100)(10)dl = 32{,}000 \; dl$$

Example 2 (Alternative Method)
Convert 32 hectograms = _____ decigrams.

$$32 \; hg \; \times \; \frac{10 \; dag}{1 \; hg} \times \frac{10 \; g}{1 \; dag} \times \frac{10 \; dg}{1 \; g} = 32(10)(10)(10) \; dg = 32{,}000 \; dg$$

Because each step in the alternative method involved multiplying by 10, we were able to use the shortcut of "moving" the decimal place by one or by adding a zero for each conversion step.

In this lesson we will begin converting from smaller units to larger units, and we will see how this works with both methods and the shortcut.

Figure 1

1 **kilo**unit	1 **hecto**unit	1 **deka**unit	1 unit	10 **deci**units	100 **centi**units	1,000 **milli**units
1,000 unit	100 unit	10 unit	1 unit	1 unit	1 unit	1 unit

The standard method of multiplying by a conversion factor from this chart will not be affected by moving from smaller units to larger units. As before, you just need to be sure that when you multiply by the conversion ratio, the top contains the unit you are converting to, and the bottom contains the unit you are converting from. This way all the units will cancel out except the one you want in the end. Note that some of the numbers in this lesson will be in decimal form.

Example 3

Convert _____ kilograms = 35 grams

$$35 \text{ g} \times \frac{1 \text{ kg}}{1,000 \text{ g}} = 0.035 \text{ kg}$$

<u>0.035</u> kilograms = 35 grams

Example 4

Convert _____ kilometers = 670 centimeters

$$670 \text{ cm} \times \frac{1 \text{ m}}{100 \text{ cm}} \times \frac{1 \text{ km}}{1,000 \text{ m}} = 0.0067 \text{ km}$$

<u>0.0067</u> kilometers = 670 centimeters

The alternative method you used to convert larger units to smaller units was to multiply by a factor of 10 for each step, which moved the number that many places to the left (or added that many zeroes). Now that you are multiplying by a factor of 1/10, this is reversed, and the number moves in the other direction.

Example 5

Convert 3.9 dekaliters = _____ milliliters

$$3 \text{ dal} \times \frac{10 \text{ d}}{1 \text{ dal}} \times \frac{10 \text{ ml}}{1 \text{ d}} = 3(10)(10)\text{ml} = 300 \text{ ml}$$

Shortcut: The decimal moved 2 places to the right. 3.00 → 300.
3.9 dekaliters = <u>300</u> milliliters

Example 6
Convert 280 centimeters = _____ hectometers

$$280 \text{ cm} \times \frac{1 \text{ dm}}{10 \text{ cm}} \times \frac{1 \text{ m}}{10 \text{ dm}} \times \frac{1 \text{ dam}}{10 \text{ m}} \times \frac{1 \text{ hm}}{10 \text{ dam}}$$

$$= 280 \left(\frac{1}{10}\right)\left(\frac{1}{10}\right)\left(\frac{1}{10}\right)\left(\frac{1}{10}\right) \text{hm} = 0.028 \text{ hm}$$

Shortcut: Decimal moved 4 places to the left. 00280. → 0.0280
280 centimeters = <u>0.028</u> hectometers

Notice that the decimal point appeared to move one place to the right for each step from larger to smaller (10/1 ratios) but appeared to move one place to the left for each step from smaller to larger (1/10 ratios). On the other hand, the number itself was moving one place to the left for each step from larger to smaller and one place to the right for each step from smaller to larger.

Because it can be difficult when using the shortcut to remember whether the number or the decimal point should be moving to the left or to the right, it can help to think of familiar situations from your experience, like changing money from smaller to larger units and vice versa.

It helps that American money uses a modified form of the same metric prefixes for tenths and hundredths, where instead of "decidollars" we have "dimes" and instead of "centidollars" we have "cents." As you know from experience, going from dollars to cents increases the number of individual units, but going from cents to dollars decreases the number of individual units.

Example 7
Convert 5 dollars = _____ cents

Estimate: There should be more cents than dollars.

$$5 \text{ dollars} \times \frac{10 \text{ dimes}}{1 \text{ dollar}} \times \frac{10 \text{ cents}}{1 \text{ dime}} = 5(10)(10)\text{cents} = 500 \text{ cents}$$

Shortcut: The decimal moved 2 places to the right. 5.00 → 500.
5 dollars = <u>500</u> cents

Example 8

Convert _____ dollars = 700 cents

Estimate: There should be fewer dollars than cents.

$$700 \cancel{\text{cents}} \times \frac{1 \cancel{\text{dime}}}{10 \cancel{\text{cents}}} \times \frac{1 \text{ dollar}}{10 \cancel{\text{dimes}}}$$

$$= 700 \left(\frac{1}{10}\right)\left(\frac{1}{10}\right) \text{dollars} = 7 \text{ dollars}$$

Shortcut: The decimal moved 2 places to the left. 700. → 7.00
<u>7</u> dollars = 700 cents

Example 9

Convert _____ hectograms = 560 centigrams

Estimate: There should be more centigrams than hectograms.
The decimal moved 5 places to the left. 000560. → 0.00560

<u>0.0056</u> hectograms = 560 centigrams

Example 10

Convert 1.05 kilograms = _____ decigrams

Estimate: There should be fewer kilograms than decigrams.
The decimal moved 4 places to the right. 001.05 → 10,500

1.05 kilograms = <u>10,500</u> decigrams

Computing Area and Circumference
of a Circle with Pi ≈ 3.14

Computing the Circumference

The formula for the circumference of a circle is πd or 2πr. Rectangles have area and perimeter, while circles have area and circumference. The *circumference* is the distance around the outside of a circle. The *diameter* is the length of a line segment that passes through the center and that touches both edges of a circle. ("Meter" means measure, and "dia" means through). Pi or π is the ratio of the circumference to the diameter. We use the Greek letter to represent this ratio because it is a number that cannot be written as a fraction, and if we try to write it as a decimal, it continues infinitely without repeating.

$$\pi = 3.14159265358979323846264338\ldots$$

For convenience, we often use a number that is very close to π instead. This lets us calculate a circumference which isn't exact but which can be written out as a fraction or as a decimal. One number that is very close to π is 22/7 (which as a decimal is $3.14\overline{285714}$). Another is 3.14. We will be using 3.14 as an approximation for π in this book because it gives us practice multiplying decimals.

Since π is the ratio between the circumference of a circle and the diameter, the circumference by definition is equal to the diameter times π, which we can write as πd. When we know the diameter of a circle we can use this formula to find the circumference, as in example 1.

Example 1
Find the circumference of the circle.

Circumference = πd
Circumference ≈ (3.14)(10 in)
Circumference ≈ 31.4 in

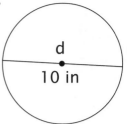

The *radius* is the distance from the center of the circle to the edge of the circle itself. To understand the more common formula for the circumference, 2πr (where "r" represents the radius of the circle), you can change the order of the three factors to π(2r). Since d (for diameter) is the same as twice the radius, then πd = π(2r). We can then rearrange the formula to 2πr, as used in the next two examples.

Example 2
Find the circumference of the circle.

Circumference = 2πr
Circumference ≈ 2(3.14)(2.5 ft)
Circumference ≈ 15.7 ft

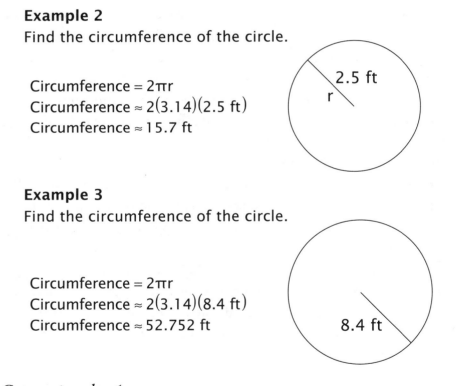

Example 3
Find the circumference of the circle.

Circumference = 2πr
Circumference ≈ 2(3.14)(8.4 ft)
Circumference ≈ 52.752 ft

Computing the Area

The formula for the area of a circle is πr^2. The symbol π and "r" have the same meanings as before, and the exponent "2" tells us that the radius is being used as a factor twice. The reason for this formula can be a little more difficult to visualize.

The following diagram may help you compare the area of a circle to the area of a square whose sides have the same length as the radius of the circle.

Figure 1

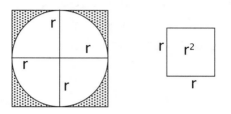

Notice that the number of squares with a side length of r and an area of r^2 which can combine to equal the area of the circle has to be some number between three and four. The actual number of squares that would cover the same area as the circle is π.

Example 4
Find the area of the circle.

$$\text{Area} = \pi r^2$$
$$\text{Area} \approx 3.14(4.5 \text{ in})^2$$
$$\text{Area} \approx 3.14(4.5 \text{ in})(4.5 \text{ in})$$
$$\text{Area} \approx 3.14(4.5)(4.5)(\text{in})(\text{in})$$
$$\text{Area} \approx 3.14(20.25)(\text{in}^2)$$
$$\text{Area} \approx 63.585 \text{ in}^2$$

Since these problems include measurement units, we are multiplying the units as well as the numbers. Think of 3 yd × 3 yd as 3 × yd × 3 × yd. We can rewrite this as 3 × 3 × yd × yd = 9 yd². The exponent shows that we used the value of a yard as a factor two times.

Including units with your measurement calculations can help you see when units should be squared in your answer. For example, 3 cm × 4 cm = 12 cm² because we are multiplying the centimeters as well as the numbers. If we know the area of a rectangle and the length of one side, and we want to find the length of the other side, we would write it as 12 cm² ÷ 4 cm = 3 cm. Dividing cm² by cm gives us just cm.

Example 5

Find the area of the circle.

Since the diameter is given, divide it by two to find the radius.

$d = 2r$, thus 13.6 ft = $2r$, and 6.8 ft = r

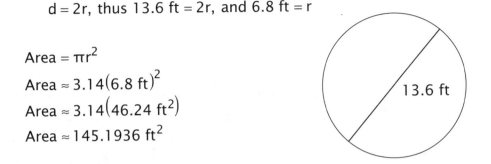

Area = πr^2

Area $\approx 3.14(6.8 \text{ ft})^2$

Area $\approx 3.14(46.24 \text{ ft}^2)$

Area $\approx 145.1936 \text{ ft}^2$

More on Perimeter and Area

The circumference of a circle corresponds to the perimeter of figures such as rectangles, squares, parallelograms, and triangles. Perimeter was taught in earlier levels of Math-U-See and is reviewed in the *Zeta* student book in the Quick Reviews for lessons 16, 17, and 18. Finding the area of figures other than circles was also taught in previous levels. Area is reviewed in lessons 19, 20, and 21 in the *Zeta* student book.

Introduction to Negative Numbers

Operations with **negative numbers** will be taught in detail in *Pre-Algebra*. However, students may already be encountering negative numbers in real-world contexts. Examples are temperatures above and below zero, elevations above and below sea level, positive and negative electrical charges, and credits and debits on a bank account.

Starting with lesson 15, several of the Application and Enrichment pages discuss negative numbers. It is also helpful to look for the applications mentioned above and discuss the meaning of the negative numbers involved. For example, –100 feet denotes a negative direction from sea level. However, if you dive to a depth of –100 feet, you will travel a distance of 100 feet, which is the absolute value of –100 feet.

The *absolute value* of a number is always a positive value that describes distance rather than direction. It is indicated by two vertical lines surrounding a number.

$$|\,4\,| = 4 \text{ and } |-4\,| = 4.$$

Dividing a Decimal by a Whole Number

As for most math topics, there is "how to do it" and "why you do it." When dividing a decimal by a whole number, put the decimal point directly above the decimal point inside the "house," and then divide the numbers. The reason for this is related to multiplication. The missing factor on top of the house (known as the *quotient*), times the known factor (*divisor*) equals the product inside the house (*dividend*). Example 1 can be read as, "What number times 3 equals 1.2?" The answer, "0.4," is a decimal. We could also read it as "Three times what number equals 1.2?" We know that $0.4 + 0.4 + 0.4 = 1.2$, so $3 \times (0.4) = 1.2$. Another option is to focus on relative place value, like we have been doing with multiplication. The number 1.2 divided by 3 is 0.4, and 12 divided by 3 is 4. We can think about the problem as 12 divided by 3, and then make sure that the answer is written in the same place (tenths place) as the number we started with. See example 1.

Figures 1 and 2 show how multiplication and division are related.

Figure 1

$$\begin{array}{r} \text{factor} \\ \text{factor} \overline{\smash{\big)}\ \text{product}} \end{array}$$

Figure 2

$$\begin{array}{r} \text{quotient} \\ \text{divisor} \overline{\smash{\big)}\ \text{dividend}} \end{array}$$

Example 1

$$\downarrow$$

$$3 \overline{\smash{\big)}\ 1.2} \;\rightarrow\; 3 \overline{\smash{\big)}\ 1\overset{.}{.}2} \;\rightarrow\; 3 \overline{\smash{\big)}\ 1\overset{.4}{.}2}$$

Always check your work: $3 \times (0.4) = 1.2$

Example 2

$$3\overline{)\smash{.12}} \rightarrow 3\overline{)\overset{.}{.12}} \rightarrow 3\overline{)\overset{.04}{.12}}$$

Always check your work: $3 \times (0.04) = 0.12$

In the examples, notice that the division isn't difficult. It is where to put the places—or where to place the decimal point—that is tricky. Put the decimal point directly above the decimal point in the dividend, and then begin putting the missing factor directly above the number inside the "house." The four is directly above the two in both examples, and the decimal points are directly above each other. However, in example 2 the "12" starts in the hundredths place, so the "4" also goes in the hundredths place, and we put a zero in the tenths place.

Example 3

Estimate: Seven is close to eight, which is a multiple of four, so think, "What times four is equal to eight?" The answer is two, so seven divided by four is almost two.

```
      1.78        Check:
  4 ) 7.12            4                    1.78
     -4           × 1.78                 ×   4
      31             .32      or          7.12
     -28            2.8
      32            4.
     -32            7.12
```

The fraction 1/8 is the same as $1 \div 8$. The number 1 is the same as 1.0 or 1.000. You can add zeros to a number without changing its value as long as they are not in between the decimal point and a digit. The number 100.0 is not the same as 1.000. In example 4 we are going to change a fraction into a decimal by dividing one by eight.

Example 4

Estimate: Eight is close to 10. Think, "What times 10 is equal to one?" The answer is one tenth or 0.1, so try 0.1 as the first step.

```
          .125      Check:
      ┌─────────
   8 │ 1.000          8
      −8            × .125                .125
      ───            ──────                × 8
      20             .040      or       ──────
     −16             .16               1.000
     ───             .8
      40           ──────
     −40           1.000
```

Example 5

Estimate: 8.06 is close to 10. Think, "What times five is equal to 10?" The answer is two, so 8.06 divided by five is almost two.

```
         1.612       Check:
      ┌─────────
   5 │ 8.060            5
      −5             × 1.612
      ───            ───────
      3 0             .010             1.612
     −3 0             .05      or       × 5
     ────             3.0             ───────
       06             5.0             8.060
      −05           ───────
      ───            8.060
        10
       −10
```

In example 5, we again needed to add a zero in order to finish solving the problem. Adding a zero at the end of the given digits *after* the decimal place is fine. If we added a zero in between the given digits, such as between the 8 and the 6, the places of the digits would change, which would change the meaning of the numbers and would lead to an incorrect answer.

Example 6

Estimate: 200.07 is close to 200, and 9 is almost 10, so think, "What times 10 is equal to 200?" The answer is 20, so try 2 as the first step.

```
         22.23
    9 ) 200.07        Check:
       −18                  9
        20            × 22.23
       −18                .27      or        22.23
        20               1.8               ×    9
       −18              18.                200.07
        27             180.
       −27            200.07
```

Dividing a Whole Number by a Decimal
and Word Problems

In this lesson, we will learn a procedure for dividing a whole number by a decimal. Let's think it through using money for an example, and we will see why this procedure works.

In example 1, the question is, "How many quarters of a dollar ($0.25), or how many groups of 25¢, are there in $6.00?" Remember that $0.25 is a decimal, which is a fraction. Since we are asking how many fractions are in a whole number, the answer should be larger than the divisor.

Example 1

$$\overset{24}{\$0.25 \overline{)\ \$6}}$$

If there are four quarters in one dollar, then there must be four times six, or 24, quarters in six dollars.

The procedure for dividing a whole number by a decimal is to make the divisor (the number outside the box) a whole number so that you are dividing a number by a whole number. To make 0.25 a whole number, we need to multiply it by 100, and 100×0.25 is 25. If we only multiplied the divisor only by 10, it would be 2.5, which is still a decimal. Recall that when we multiply the divisor by 100, we need to multiply the dividend by the same number as well, and 100 times 6 is 600. See example 2 on the next page.

Example 2

$$25 \overline{)600}$$
$$24$$
$$-50$$
$$100$$
$$-100$$

$0.25 \times 100 = 25$ and $6 \times 100 = 600$

$$\frac{\$6}{\$0.25} \times \frac{100}{100} = \frac{600}{25}$$

Notice that this process is really the same as making an equivalent fraction. Multiplying the divisor (denominator) and dividend (numerator) by the same number doesn't change the quotient. When we multiply both numbers by a multiple of 10, whether 10, 100, 1,000, etc., the decimal points appear to move the same number of spaces. No matter how many spaces we move the decimal point to make the divisor (on the outside) a whole number, we move the decimal point on the inside (the dividend) the same number of places. See example 3.

Example 3

$$.25 \overline{)6.00} \rightarrow .25 \overline{)6.00} \rightarrow 25 \overline{)600}$$

$$24.$$
$$-50$$
$$100$$
$$-100$$

$$\begin{array}{r} .25 \\ \times\ 24 \\ \hline 1.00 \\ 5.0 \\ \hline 6.00 \end{array}$$

In example 4 the question is, "How many sets of three dimes (0.3 or 30¢)are there in 18 dollars ($18.00)?" Remember that 0.3 and 0.30 are fractions written a different way. Since you are asking how many fractions are in a whole number, the answer should be larger than 18.

Example 4

$$.3 \overline{)18}$$

There are a little more than three sets of $0.30 in one dollar, so there must be at least three times 18, or 54 sets, in 18 dollars.

$$.3\overline{)18} \rightarrow .3\overline{)18.0} \rightarrow 3.\overline{)180}$$

$$\begin{array}{r} 60. \\ 3.\overline{)180} \\ \underline{-18} \\ 00 \\ \underline{-00} \end{array}$$

$$\begin{array}{r} 60 \\ \times\ .3 \\ \hline 18.0 \end{array}$$

60 is greater than 54, so it fits with our estimate of 54.

In example 5, the question is, "How many sets of five hundredths of a dollar ($0.05 or 5¢) are there in 121 dollars ($121.00)?" Remember that $0.05 is a fraction of a dollar.

Example 5

Since there are 20 nickels in a dollar, the answer should be 20 times larger than 121.

$$.05\overline{)121} \rightarrow .05\overline{)121.00} \rightarrow 5.\overline{)12,100.}$$

$$\begin{array}{r} 2,420. \\ 5.\overline{)12,100.} \\ \underline{-10} \\ 2\ 1 \\ \underline{-2\ 0} \\ 10 \\ \underline{-10} \\ 00 \\ \underline{-0} \\ 0 \end{array}$$

20 times 121 is 2,420. Our estimate was correct.

Word Problems

Here are a few more multi-step word problems to try. You may want to read and discuss these with your student as you work out the solutions together. Again, the purpose is to stretch, not to frustrate. If you do not think the student is ready, you may want to come back to these later.

The answers are at the end of the solutions at the back of this book.

1. The boys ordered several pizzas for the weekend. When the first evening was over, the following amounts of pizza were left over: one-fourth of the pepperoni pizza, one-half of the cheese pizza, three-fourths of the mushroom pizza, and one-fourth of the supreme pizza. The next morning, each boy ate the equivalent of one-fourth of a pizza for breakfast. If that finished the pizza, how many boys were there?

2. Dan read that an average snowfall of 10 inches yields one inch of water when melted. He also found that five inches of very wet snow or 20 inches of very dry snow will both melt to an inch of water. He made measurements for a storm that started with 5.3 inches of average snowfall. The precipitation changed to wet snow and dropped another 4.1 inches. The weather continued to warm up, and the storm finished with 1.5 inches of rain. About how much water fell during the storm? Round your answer to tenths.

3. Jim bought edging to go around a circular garden with a radius of three feet. Later he decided to double the diameter of the garden. How many more feet of edging must he buy?

4. One packet of flower seeds was enough to just fill the area of the smaller garden in #3. How many packets of seed are needed for the larger garden? Round the area of each garden to the nearest square foot before solving the problem.

Solving for an Unknown

An *algebraic expression* is a combination of one or more numbers, symbols representing numbers, and operation symbols. For example, X + 5 is a simple algebraic expression which has two *terms*. Just as words and phrases are combined to make a sentence, algebraic expressions may be combined to make equations. In previous levels of Math-U-See, we solved a number of different kinds of equations. In this book, we are going to solve equations that include decimals.

When solving for the unknown in an equation such as 0.3X = 12, we need to get the unknown (in this case X) by itself. We want our final step to be X = ___ with a number in the blank. A *coefficient* means a factor within a term. Usually we use this word when we want to talk about the number that a variable or unknown is being multiplied by. In this case 0.3 is the coefficient of X. We need to make the coefficient 1, which we can also think of as having no written coefficient because 1 times X is X.

Since 0.3X means X multiplied by 0.3, we can multiply by the reciprocal, or we can use the inverse of multiplication, which is division, to make the coefficient of the unknown 1. See example 1 for our solution.

Example 1
Divide both sides by 0.3, so the left side will be 1 · X or X without a written coefficient.

$$3X = 12$$
$$\frac{0.3X}{0.3} = \frac{12}{.3} \rightarrow 0.3\overline{)12\smile} \rightarrow 3.\overline{)120.}^{40.}$$
$$X = 40$$

Check the answer.

$$0.3(40) = 12$$
$$12 = 12 \text{ It works.}$$

Example 2

Divide both sides by 0.05, so the left side will be $1 \cdot R$ or R without a written coefficient.

$$0.05R = 6$$
$$\frac{0.05R}{0.05} = \frac{6}{0.05} \rightarrow 0.05\overline{)6} \rightarrow 5.\overline{)600.}^{120.}$$
$$R = 120$$

Check the answer.

$$0.05(120) = 6$$
$$6 = 6 \text{ It works.}$$

Dividing a Decimal by a Decimal

Before I tell you how to do these problems, let's do one just by thinking. We'll use money for our example to make it clearer. See example 1. The question is, "How many nickels are there in 75 cents?"

Example 1

$$\overset{\displaystyle 15.}{.05\,\overline{\smash{\big)}\,.75}}$$

From what we know of money, there are 15 nickels in 75¢.

$$15 \times (\$0.05) = \$0.75$$

To do these problems you can first make the divisor, or the known factor, a whole number. To make 0.05 a whole number, we have to multiply it by 100. That makes it the whole number 5. Remember that we need to multiply *both* the divisor and the dividend by 100, or it changes the whole problem. Multiplying 0.75 times 100 gives 75. Our old problem in new clothes is 75 ÷ 5, as shown in example 2. We get the same answer as we did in example 1.

Example 2

$$\begin{array}{r} 15 \\ 5\,\overline{\smash{\big)}\,75} \\ \underline{-5} \\ 25 \\ \underline{-25} \end{array}$$

When we multiply both numbers by the same multiple of 10, the decimal points appear to move the same number of spaces in both numbers. So we could say that no matter how many spaces we have to move the decimal point to make the divisor (on the outside) a whole number, we move the decimal point on the inside (the dividend) the same number of places. See the following example. The number 3 is the same as 3.0.

Example 3

$$.03 \overline{) 1.974} \quad \rightarrow \quad .03 \overline{) 1.974} \quad \rightarrow \quad 3 \overline{) 197.4}$$

$$
\begin{array}{r}
65.8 \\
3 \overline{) 197.4} \\
-18 \\
\hline
17 \\
-15 \\
\hline
2\,4 \\
-2\,4 \\
\hline
\end{array}
$$

$$
\begin{array}{r}
.03 \\
\times\ 65.8 \\
\hline
0.024 \\
0.15 \\
1.80 \\
\hline
1.974 \\
\end{array}
\qquad
\begin{array}{r}
65.8 \\
\times\ .03 \\
\hline
1.974 \\
\end{array}
$$

Notice that in example 4 we need to add a zero to the dividend because there aren't enough places to complete the division.

Example 4

$$.006 \overline{) 63.87} \quad \rightarrow \quad .006 \overline{) 63.87} \quad \rightarrow \quad 6 \overline{) 63870}$$

$$
\begin{array}{r}
10645 \\
6 \overline{) 63870} \\
-6 \\
\hline
3 \\
-0 \\
\hline
38 \\
-36 \\
\hline
27 \\
-24 \\
\hline
30 \\
-30 \\
\hline
\end{array}
$$

$$
\begin{array}{r}
10645 \\
\times\ .006 \\
\hline
63.870 \\
\end{array}
$$

LESSON 21

Decimal Remainders

Remainders in a division problem involving decimals can be tricky. There are four possibilities, and I will go through each one with an accompanying example.

Option 1

With certain single-digit divisors, such as 2, 4, 5, or 8, if you add enough zeros to the end of the dividend, eventually you will get the answer, or quotient, without a remainder.

Example 1
Divide 2.3 by 8.

```
        .2875
    8 ⟌ 2.3000
       −1 6
          70
         −64
          60
         −56
           40
          −40
```

Option 2

With other single-digit divisors, such as 3, 6, 7, or 9, sometimes you can add all the zeros you want and never arrive at an answer without a remainder. It is with those problems in mind that we address options 2, 3, and 4. The most-used option is to select a place value that fits your needs and round to that value. For example, we often decide that we really don't need to know more than the hundredths place. Before you begin dividing in a case like this, decide what place you want the answer rounded to. This could be thousandths or millionths or any other place value, but for example 2, hundredths was chosen. All we need to do is divide to the thousandths place and then round to the closest value in the hundredths place.

Example 2
Divide 29.6 by 7.

$$
\begin{array}{r}
4.228 \\
7{\overline{\smash{\big)}\,29.600}} \\
\underline{-28} \\
1\,6 \\
\underline{-1\,4} \\
20 \\
\underline{-14} \\
60 \\
\underline{-56}
\end{array}
$$

4.228 rounds to 4.23

Option 3

Another option for divisors such as 3, 6, 7, or 9 is to look for a pattern. We call decimals with a pattern *repeating decimals.* When you divide by 3, eventually you will start seeing 3s in the quotient that go on and on. When you find a pattern like this, put a line above the repeating part to indicate that it goes on and on. Other divisors like 7 have very long numbers in their patterns before they repeat, as in example 3C. Study examples A, B, and C.

Example 3A
Divide 8.5 by 3.
2.833 . . . equals $2.8\overline{3}$

$$
\begin{array}{r}
2.833 \\
3{\overline{\smash{\big)}\,8.500}} \\
\underline{-6} \\
2\,5 \\
\underline{-2\,4} \\
10 \\
\underline{-9} \\
10 \\
\underline{-9}
\end{array}
$$

Example 3B

Divide 4.6 by 9.

5.111 . . . equals 5.$\overline{1}$

```
       .5111
    9 ⟌ 4.6000
      −4 5
         10
         −9
         10
         −9
         10
         −9
```

Example 3C

Divide 9.5 by 7.

```
        1.3571428
```

The digits 571428 repeat.

Do you see the repeat starting
again, beginning with 57?

```
        1.357142857
     7 ⟌ 9.500000000
       −7
        2 5
       −2 1
          40
         −35
          50
         −49
          10
          −7
          30
         −28
          20
         −14
          60
         −56
          40
         −35
          50
         −49
```

Option 4

Occasionally it is useful to express a decimal answer as a percentage with a fraction in the last place. For example, we might say that 33⅓ percent of the people surveyed liked a certain product. In example 4, we divide to the hundredths place and want to make the thousandths place a fraction. So we put the 4 on top of the 7 because the next step in the problem is 4 divided by 7, and leave it like that. The answer is 0.27 $\frac{4}{7}$ or 27 $\frac{4}{7}$%.

Remember though, that the 4/7 is really 0.04 divided by 7, so it represents 4/7 of a hundredth, not 4/7 of one. When we subtract 14 from 1.93, the first number is really 1.4. The 53 is really 0.53 if we leave the decimal points in. When you check the problem, 7 times 0.27 is 1.89, and 7 times 0.04/7 is 0.04. Then 0.04 added to 1.89 is 1.93. Think of the 4/7 as occupying the hundredths place along with the 7 in 0.27.

Example 4
Divide 1.93 by 7.

$$27\tfrac{4}{7}\%$$

$$
\begin{array}{r}
.27 \;\; \frac{4}{7} \\
7\overline{)\,1.93} \\
\underline{-1\,4} \\
53 \\
\underline{-49} \\
4
\end{array}
$$

LESSON 22

More Solving for an Unknown

Now we learn how to solve slightly more complex algebraic equations. When asked to solve for the unknown X in 0.3X + 0.2 = 2, we need to get X by itself. The final step should be X = ___ with a number in the blank. We have learned how to make the coefficient 1. Now we have to learn how to make the number added to 0.3X a zero. In the example, the number being added is 0.2.

We want the *variable,* or unknown, to be alone on one side of the equation and everything else to be on the other side of the equation. The opposite, or inverse, of adding 0.2 is subtracting 0.2 from both sides. In example 1, we'll get 0.3X by itself, and then we'll divide by 0.3 to make our variable just X.

Example 1

Subtract 0.2 from both sides so that the left side has just the variable.

Divide both sides by 0.3 so that the left side will be 1 · X or just X.

$$3X + 0.2 = 2$$
$$\underline{-0.2 \quad -0.2}$$
$$0.3X + 0 = 1.8$$
$$0.3X = 1.8$$
$$\frac{0.3X}{0.3} = \frac{1.8}{0.3}$$
$$X = 6$$

$$\rightarrow 0.3\overline{)18} \rightarrow 3.\overline{)18.}^{\,6.}$$

Check the answer.

$$0.3(6) + 0.2 = 2$$
$$1.8 + 0.2 = 2$$
$$2 = 2 \quad \text{It works.}$$

Example 2

Subtract 3.3 from both sides so that the left side has just the variable.

$$1.5J + 3.3 = 13.8$$
$$\underline{-3.3 \quad -3.3}$$
$$1.5J + 0 = 10.5$$
$$1.5J = 10.5$$

Divide both sides by 1.5 so that the left side will be $1 \cdot J$ or just J.

$$\frac{1.5J}{1.5} = \frac{10.5}{1.5} \rightarrow 1.5\overline{)10.5} \rightarrow 15\overline{)105.}^{7.}$$
$$J = 7$$

Check the answer.

$$1.5(7) + 3.3 = 13.8$$
$$10.5 + 3.3 = 13.8$$
$$13.8 = 13.8 \text{ It works.}$$

Example 3

$$0.12A + 0.052 = 0.1$$

Subtract 0.052 from both sides.

$$\underline{-0.052 \quad -0.052}$$
$$0.12A + 0 = 0.048$$
$$0.12A = 0.048$$

Divide both sides by 0.12.

$$\frac{0.12A}{0.12} = \frac{0.048}{0.12} \rightarrow 0.12\overline{)0.048} \rightarrow 12\overline{)4.8}^{0.4}$$
$$A = 0.4$$

Check the answer.

$$0.12(0.4) + 0.052 = 0.1$$
$$0.048 + 0.052 = 0.1$$
$$0.1 = 0.1 \text{ It works}$$

Volume

Finding the volume of cubes and other rectangular solids was taught in previous levels of Math-U-See. You can find review of these concepts in the *Zeta* student book in the Quick Reviews for lessons 22 and 23. A *rectangular prism* is another name for a rectangular solid.

Transform Any Fraction
to a Decimal and a Percentage

Recall that the line separating a fraction into a numerator and denominator also means "divided by." The fraction 1/4 means one divided by four. Now that we know how to divide numbers such as these, we can change any fraction into a decimal with hundredths. When dividing to make a percentage, always stop at the hundredths place because that is what percentages are—another way of writing hundredths. In examples 2–4, in order to make the answer stop at hundredths, we chose to write the thousandths place as a fraction. If we had rounded the answer it would be less precise. Once we have changed to hundredths, we can write it as a percentage. For more on final remainders, see lesson 21.

Example 1
Transform 1/4 into a decimal and a percentage.

$$\frac{1}{4} = 4\overline{\smash{\big)}1.00} \begin{array}{r} .25 \\ \hline -8 \\ \hline 20 \\ -20 \end{array} = 25\%$$

Example 2
Transform 2/3 into a decimal and a percentage.

$$\frac{2}{3} = 3\overline{\smash{\big)}2.00} \begin{array}{r} .66 \\ \hline -1\,8 \\ \hline 20 \\ -18 \\ \hline 2 \end{array} = 0.\overline{6} \text{ or } 0.66\frac{2}{3} = 66\frac{2}{3}\%$$

Example 3
Transform 3/7 into a decimal and a percentage.

$$\frac{3}{7} = 7\overline{)\begin{array}{r} .42 \\ 3.00 \\ \end{array}} \quad = 0.42\frac{6}{7} = 42\frac{6}{7}\%$$

$$\begin{array}{r} -2\ 8 \\ \hline 20 \\ -14 \\ \hline 6 \end{array}$$

Example 4
Transform 5/8 into a decimal and a percentage.

$$\frac{5}{8} = 8\overline{)\begin{array}{r} .62 \\ 5.00 \\ \end{array}} \quad = 0.62\frac{4}{8} = 0.62\frac{1}{2} = 62\frac{1}{2}\%$$

$$\begin{array}{r} -4\ 8 \\ \hline 20 \\ -16 \\ \hline 4 \end{array}$$

Example 5
Transform 1/3 into a decimal and a percentage.

$$\frac{1}{3} = 3\overline{)\begin{array}{r} .33 \\ 1.00 \\ \end{array}} \quad = 0.33\frac{1}{3} = 33\frac{1}{3}\%$$

$$\begin{array}{r} -9 \\ \hline 10 \\ -9 \\ \hline 1 \end{array}$$

Earlier in the book, the student should have memorized several fractions the respective percentages for the fractions 1/2, 1/4, 3/4, and 1/5. Of course, once you know one-fifth you can easily find the other fifths by multiplying by their numerator. So since one-fifth is 20%, two-fifths is 40%, etc.

Figure 1 lists other fractions that I suggest you also memorize. They have helped me in everyday situations for years. You may find that in your experience, other fractions are helpful to memorize as well.

Figure 1

$$\frac{1}{3} = 33\frac{1}{3}\% \qquad\qquad \frac{1}{8} = 12\frac{1}{2}\%$$

$$\frac{2}{3} = 66\frac{2}{3}\% \qquad\qquad \frac{1}{9} = 11\frac{1}{9}\%$$

$$\frac{1}{7} = 14\frac{2}{7}\% \qquad\qquad \frac{1}{20} = 5\%$$

I usually use these equivalents when I can round them, since I am finding an approximate answer. So I think of one-third as 33%, two-thirds as 67%, one-ninth as 11%, etc. If I know one-ninth is approximately 11%, then five-ninths is five times that or about 55%. The more you have tucked away upstairs (in your memory), the more you will have readily available to solve problems that arise unexpectedly. The more math you know, the more you can and will use.

A good application for transforming a fraction into a percentage is finding your percentage score when grading a test. First write a fraction showing how many are correct. If you get 18 right out of 20, your score is 18/20 or 9/100 or 90%.

Decimals as Rational Numbers
and Word Problems

A *rational number* is a number that may be expressed as the result of dividing two numbers. A fraction, such as 7/8, is a rational number. Whole numbers may also be expressed as rational numbers, such as $5 = \frac{5}{1}$. Since decimals are fractions written in the base 10 system, they are a type of rational number. For example, to write the decimal 0.23 as a fraction, simply change it to 23/100. The numbers 0.23 and 23/100 are identical in value. One is written as a fraction while the other is written in the base 10 system as a decimal. For help in learning what a rational number is, notice that the first five letters in R-A-T-I-O-nal refer to ratios.

Most of our study of ratios has been focused on fractions, but there were several application and enrichment pages (starting with 11G) that examined other types of ratios. For example, think of a classroom. If there are 10 boys and 15 girls, the ratio of boys to girls is 2 to 3, which can be written as 2/3. Since fractions show the ratio of the specific number of parts (numerator) to the number of those parts in a whole unit (denominator), fractions are a special type of ratio. Therefore, we can refer to fractions — and any other number that could be written as a fraction — as rational numbers. In the previous lesson, we transformed fractions into decimals and into percentages. In this lesson, we will be doing the reverse: transforming decimals into reduced fractions.

Example 1
Transform 0.29 into a reduced fraction.

$$0.29 = \frac{29}{100}$$ You can't reduce this fraction any further, so it is the final answer.

Example 2

Transform 0.8 into a reduced fraction.

$$0.8 = \frac{8 \div 2}{10 \div 2} = \frac{4}{5}$$

Example 3

Transform 0.036 into a reduced fraction.

$$0.036 = \frac{36 \div 4}{1000 \div 4} = \frac{9}{250}$$

Example 4

Transform 0.625 into a reduced fraction.

$$0.625 = \frac{625 \div 25}{1000 \div 25} = \frac{25 \div 5}{40 \div 5} = \frac{5}{8}$$

Example 5

Transform 0.035 into a reduced fraction.

$$0.035 = \frac{35 \div 5}{1000 \div 5} = \frac{7}{200}$$

Word Problems

Here are a few more multi-step word problems to try. You may want to read and discuss these with your student as you work out the solutions together. Remember that the purpose is to stretch, not to frustrate. If you do not think the student is ready, you may want to come back to these later.

The answers are at the end of the solutions at the back of this book.

1. Emily cut two circles from a sheet of colored paper measuring 8 inches by 12 inches. One circle had a radius of 3 inches, and the other had a radius of 2.5 inches. How many square inches of paper are left over? Is it possible to cut another circle with a 3-inch radius from the paper?

2. Tom wants to buy items costing $25.35, $50.69, and $85.96. He earns $6.50 an hour doing odd jobs. If 10 percent of his income is put aside for other purposes, how many hours must he work to earn the money for his purchases? Round your answer to the nearest whole hour.

3. Three-tenths of the wooden toys were painted blue, and one-fourth of them were painted green. Half of the remaining toys were painted red, and half were painted yellow. If 300 toys are blue, how many are there of each of the other colors?

Mean, Median, and Mode

Statistics is the branch of mathematics that collects and organizes data to find patterns and make predictions or draw conclusions. Here are three terms used in statistics to describe what the average or center of a set of data is. Key letters in each word can help you remember them.

meAn = A represents Average, and we usually are referring to the mean when we say "average."

MeDian = MD represents the Middle number in the data when arranged in ascending order.

MOde = MO represents the number that occurs Most Often in the data.

Example 1

Arrange the following data in ascending order and find the mean, median, and mode: 5, 2, 5, 6, 8, 7, 9.

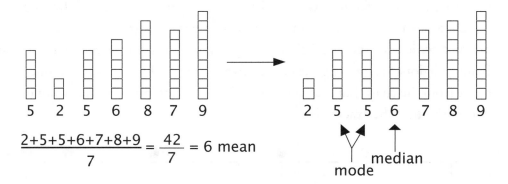

$$\frac{2+5+5+6+7+8+9}{7} = \frac{42}{7} = 6 \text{ mean}$$

In example 1, there is an odd number of data points, so the median is simply the middle point, which is six. When there is an even number of data, the median is calculated by finding the average of the two middle data. In example 2, the median is the average of seven and nine, which is eight.

Example 2
Arrange the following data in ascending order and find the mean, median, and mode: 10, 2, 9, 4, 10, 7.

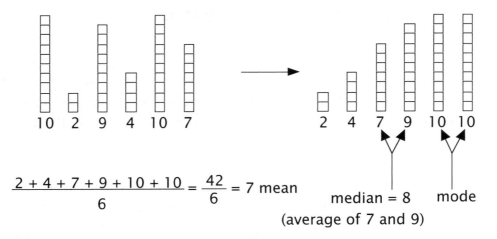

$$\frac{2 + 4 + 7 + 9 + 10 + 10}{6} = \frac{42}{6} = 7 \text{ mean}$$

median = 8
(average of 7 and 9)

mode

Statistics and Variability

Statistics are useful when a question has a number of different answers. "How tall are you?" has one correct answer. We don't need statistics to interpret the answer. On the other hand, "How tall are people in your family?" has a number of different answers. For the second question, the mean, median, and mode can be used to give you different ways to describe the center, or average, of the data.

Data can also be shown using a variety of charts and graphs. One example is a *dot plot*, which is a statistical chart that plots data on a simple number-line scale. Below is a dot plot showing the data from example 2. Notice that the mode is easy to see from the chart, but that you will still need to compute the mean and median in order to find the exact values.

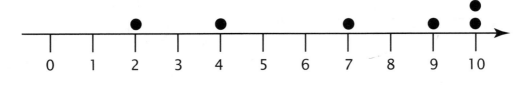

Probability

Probability is a numerical expression of how likely something is to happen or how likely a statement is to be true. It is expressed on a scale of 0 to 1, where a probability of 0 means that something is impossible, and a probability of 1 means that it is certain. Probability is expressed as a ratio between a specific outcome and of all possible outcomes:

$$\frac{\text{Specified outcome}}{\text{Possible outcomes}}$$

Example 1

You are rolling a six-sided die (singular of dice). On the sides are the numbers 1, 2, 3, 4, 5, and 6. Answer the following questions.

Question	Answer
What is the probability of rolling a 4?	$\dfrac{\text{Specified roll}}{\text{Possible rolls}} = \dfrac{4}{1-2-3-4-5-6} = \dfrac{1}{6}$
What is the probability of rolling a 5?	$\dfrac{\text{Specified roll}}{\text{Possible rolls}} = \dfrac{5}{1-2-3-4-5-6} = \dfrac{1}{6}$
What is the probability of rolling a 1, 3, or 5 (an odd number)?	$\dfrac{\text{Specified roll}}{\text{Possible rolls}} = \dfrac{1-3-5}{1-2-3-4-5-6} = \dfrac{3}{6} = \dfrac{1}{2}$
What is the probability of rolling a 2, 4, or 6 (an even number)?	$\dfrac{\text{Specified roll}}{\text{Possible rolls}} = \dfrac{2-4-6}{1-2-3-4-5-6} = \dfrac{3}{6} = \dfrac{1}{2}$
What is the probability of rolling a 1 or a 5?	$\dfrac{\text{Specified roll}}{\text{Possible rolls}} = \dfrac{1-5}{1-2-3-4-5-6} = \dfrac{2}{6} = \dfrac{1}{3}$

Example 2

You bought a book with 24 pages in it. Of the 24, 18 had printing, four had pictures, and two pages were blank. Answer the following questions.

Question	**Answer**
What is the probability of opening to a picture?	$\dfrac{\text{Specified pages}}{\text{Possible pages}} = \dfrac{4}{24} = \dfrac{1}{6}$
What is the probability of opening to printing?	$\dfrac{\text{Specified pages}}{\text{Possible pages}} = \dfrac{18}{24} = \dfrac{3}{4}$
What is the probability of opening to a blank page?	$\dfrac{\text{Specified pages}}{\text{Possible pages}} = \dfrac{2}{24} = \dfrac{1}{12}$
What is the probability of opening to pictures or printing?	$\dfrac{\text{Specified pages}}{\text{Possible pages}} = \dfrac{4+18}{24} = \dfrac{22}{24} = \dfrac{11}{12}$

Points, Lines, Rays, and Line Segments

Geometry is the study of shapes and their sizes and relative positions. In formal terms, we can say that it deals with spatial relationships. The word geometry comes from an ancient Greek term for surveying or measuring land. "Geo" means earth or land, and "metry" means measure. To measure land, we need to break it into smaller, more manageable pieces. The smallest possible piece is an imaginary figure, called a *point*. It has no measurable size, only position or location. We can't measure its width or length, so it has no dimensions and is zero-dimensional. To show something that is so small that you really can't see it, we draw a dot. The dot is the "graph" of the point and represents the point. We can label it with a capital letter. In figure 1, we call the specific point "point A."

Figure 1

·A   Point A

Infinity is a quantity that is greater than any number. It is represented by the symbol ∞. A line is a geometric figure that has infinite length and no width. It extends endlessly in two directions and (in the context of plane geometry) is perfectly straight. It is drawn with arrows on both ends. Because a line has only length, we say that it is one-dimensional.

Figures 2 and 3 show two ways to label lines. We call figure 2 "line m." In figure 3, two points (represented by capital letters) are chosen to name it "line QR" or $\overleftrightarrow{RQ}$. Note that the line symbol on top of RQ has two arrows, and that the order of the points does not matter. Also, any two points may be used.

Figure 2

$$\longleftrightarrow \quad m \quad \longrightarrow \qquad \text{Line } m$$

Figure 3

$$\longleftarrow \quad \overset{Q}{\bullet} \qquad \overset{R}{\bullet} \quad \longrightarrow \qquad \text{Line QR or } \overset{\leftrightarrow}{QR}$$

A ray may be thought of as one-half of a line. A *ray* is a geometric figure that has a specific starting point, called the endpoint or origin, at one end, and that proceeds infinitely in only one direction. If you think of a laser pointer, the pointer itself is like the origin, and the laser beam goes on infinitely. Since a ray goes in only one direction, its symbol has only one arrow. Figure 4 is labeled as $\overset{\rightarrow}{BC}$ and read as "ray BC." The order the points are written is important with rays, and the first point listed is always the origin.

Figure 4

A *line segment* is a finite piece of a line with two endpoints. You can measure a line segment. Figure 5 is labeled as $\overline{LH}$ and read as "line segment LH." The same segment could be called $\overline{HL}$ because the order of the endpoints is not important. Only endpoints can be used to name a line segment.

Figure 5

$$\overset{L}{\bullet} \underline{\hspace{8cm}} \overset{H}{\bullet} \qquad \text{Line Segment } \overline{LH}$$

Because points have no width or length, lines, rays, and line segments all contain an infinite number of points.

Planes and Symbols

A *plane* is an infinite number of connected lines lying in the same flat surface. A plane has length and width, so it has two dimensions. You can think of it as an infinite floor or tabletop except with no thickness. It is long and wide and keeps going in those two dimensions, but it is thinner than a piece of paper because it is just as thin as a line, which is as thin as a point. We can use a picture of a parallelogram to represent a plane, and we often label a plane with a lowercase letter. In figure 1, the plane is referred to as "plane b."

Figure 1

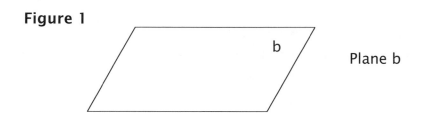

Plane b

In this lesson and the last one we are focusing on flat, two-dimensional figures that lie in the same plane. This is called *plane geometry*. Three-dimensional geometry, with length, width, and height (or depth) pertaining to space and solids is called space or solid geometry. *Solid geometry* includes shapes like cubes, cylinders, pyramids, cones, and spheres.

To recap the lessons on geometry so far, consider that we operate in a three-dimensional place called space. The room you are in has the three dimensions of length, width, and height. Look at figure 2 on the next page, and notice how we have progressed from no dimensions (a point) to one dimension (a line) to two dimensions (a plane), and finally to three dimensions.

Figure 2

point	•	zero dimensions
line	←——→	one dimension
plane	▱	two dimensions
space	⬗	three dimensions

Symbols

The symbol "~" by itself means *similar,* or the same shape but not exactly the same size. Two squares that have exactly the same shape but different measurements are said to be similar. Consider a square that is 2 inches by 2 inches and another square that is 4 miles by 4 miles. They have the same shape but are not the same size, so they are similar but not identical.

The equal sign "=" means exactly the same quantity, or *equal*, and is used if two numbers or quantities are the same. Putting the similar and equal signs together (≅) means that two shapes have exactly the same proportions and that the dimensions of the shapes have the same numerical value. We say the shapes themselves are *congruent*. Use the equal sign for quantities (including measurable quantities like the length or area of a shape). Choose the congruent sign for objects that have the same shape and size.

Figure 3

Similar	~	For shapes
Equal	=	For numbers
Congruent	≅	For shapes

LESSON 29

Angles

If two lines intersect, the opening, or space between the lines, is referred to as an *angle.* In figure 1, there are four angles shown by the arcs. These angles are named 1, 2, 3, and 4.

Figure 1

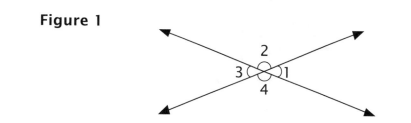

In figure 2, we are focusing on just one angle, made by drawing two rays with the same endpoint. This endpoint, or origin, is called the *vertex.* (The plural of vertex is vertices.)

Figure 2

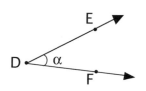

The rays are $\overrightarrow{DE}$ and $\overrightarrow{DF}$. The angle is labeled with either a number as in figure 1, or a lowercase Greek letter as in figure 2. We'll call this figure ∠α (angle alpha). Another way to identify this angle is by picking one point on each ray and the vertex to make either ∠EDF or ∠FDE. Notice that the point labeling the vertex is always in the middle.

We can measure the size of angles using degrees. Each degree is 1/360th of a full turn, or a complete circle.

A *right angle* has a measure of 90 degrees (90°), and forms a square corner. Notice that the position of the angle doesn't matter, only that its measure is 90°. Usually, a box symbol is used to indicate a right angle.

Since there are 360 degrees in a circle, a right angle is one-fourth of a circle. See figure 4.

Figure 3

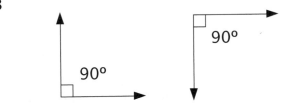

Figure 4

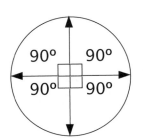

Types of Angles

An *acute angle* is greater than 0° but less than 90°.

Since they are small angles, it helps me to remember the name by thinking of "cute." Figure 1 has two acute angles.

Figure 1
0° < acute < 90°

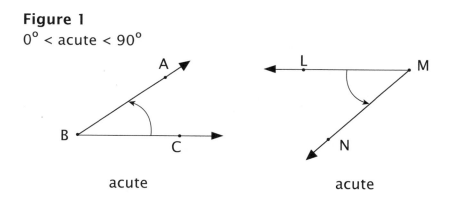

acute   acute

An *obtuse angle* is greater than 90° and less than 180°. See figure 2.

Figure 2
90° < obtuse < 180°

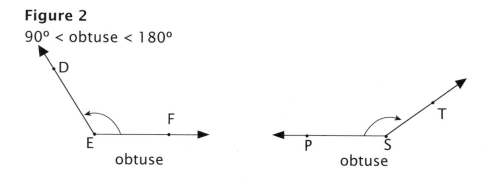

obtuse   obtuse

A *straight angle* is a lesser-known angle and is sometimes difficult to think of as an angle. It has a measure of 180°.

Figure 3

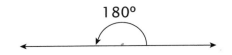

You sometimes hear of a car that skidded on ice and did a "one eighty," meaning it was going in one direction and then spun around until it was pointing in the opposite direction. This comes from the fact that it spun 180°. Figure 4 is a top view of this skid.

Figure 4

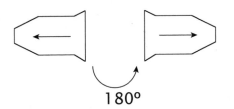

Student Solutions

Lesson Practice 1A
1. done
2. done
3. 4^2
4. 10^2
5. $5 \times 5 = 25$
6. $12 \times 12 = 144$
7. $4 \times 4 \times 4 = 64$
8. 6
9. $4 \times 4 = 16$
10. 100
11. done
12. 2^3
13. 5^2
14. 2^5
15. 8^2
16. 1^3
17. 4×4 or 16
18. 16 or 4×4

Lesson Practice 1B
1. 7^2
2. 12^2
3. 2^2
4. 6^2
5. $2 \times 2 \times 2 \times 2 = 16$
6. $1 \times 1 = 1$
7. $10 \times 10 = 100$
8. 8
9. $9 \times 9 = 81$
10. $3 \times 3 \times 3 = 27$
11. 2^1
12. 8^2
13. 2^4
14. 9^3
15. 6^4
16. 10^2
17. $5 \times 5 \times 5$ or 125
18. 125 or $5 \times 5 \times 5$

Lesson Practice 1C
1. 5^2
2. 8^2
3. 10^2
4. 1^2
5. $1 \times 1 \times 1 = 1$
6. $11 \times 11 = 121$
7. $2 \times 2 = 4$
8. $3 \times 3 \times 3 \times 3 = 81$
9. $8 \times 8 = 64$
10. $6 \times 6 = 36$
11. 7^2
12. 4^3
13. 9^2
14. 2^3
15. 7^5
16. 3^3
17. 10×10 or 100
18. 100 or 10×10

Lesson Practice 1D
1. $7 \times 7 = 49$
2. $12 \times 12 = 144$
3. $100 \times 100 = 10,000$
4. $1 \times 1 \times 1 \times 1 = 1$
5. 9
6. $3 \times 3 = 9$
7. 10^2
8. 5^3
9. 3^3
10. 4^2
11. 8^4
12. 5^3
13. $1 \times 1 \times 1$ or 1
14. 1 or $1 \times 1 \times 1$
15. done
16. $6 \div 3 = 2$
    $2 \times 1 = 2$

17. $20 \div 5 = 4$
$4 \times 4 = 16$

18. $24 \div 6 = 4$
$4 \times 5 = 20$

19. $\frac{1}{2}$ of $30 =$
$30 \div 2 = 15$
$15 \times 1 = 15$ students

20. $\frac{7}{8}$ of $16 =$
$16 \div 8 = 2$
$2 \times 7 = 14$ guests

Systematic Review 1E

1. $2 \times 2 \times 2 = 8$
2. $4 \times 4 \times 4 \times 4 = 256$
3. $11 \times 11 = 121$
4. $8 \times 8 = 64$
5. $6 \times 6 \times 6 = 216$
6. 15
7. 12^2
8. 6^2
9. 10^3
10. 1^5
11. 6^3
12. 9^2
13. 12×12 or 144
14. 144 or 12×12
15. $32 \div 8 = 4$
$4 \times 3 = 12$
16. $12 \div 6 = 2$
$2 \times 1 = 2$
17. $300 \div 3 = 100$
$100 \times 2 = 200$
18. $72 \div 4 = 18$
$18 \times 1 = 18$
19. $\frac{3}{4}$ of $600 =$
$600 \div 4 = 150$
$150 \times 3 = 450$ visitors

20. $2 \times 12 = 24$ eggs
$24 \div 6 = 4$
$4 \times 1 = 4$ eggs

Systematic Review 1F

1. $9 \times 9 = 81$
2. $13 \times 13 = 169$
3. $3 \times 3 \times 3 \times 3 \times 3 = 243$
4. $5 \times 5 \times 5 \times 5 = 625$
5. 10
6. $20 \times 20 = 400$
7. 4^2
8. 7^3
9. 8^1
10. 3^4
11. 11^2
12. 10^3
13. $7 \times 7 \times 7 \times 7$ or $2,401$
14. $2,401$ or $7 \times 7 \times 7 \times 7$
15. $55 \div 5 = 11$
$11 \times 2 = 22$
16. $210 \div 7 = 30$
$30 \times 3 = 90$
17. $90 \div 10 = 9$
$9 \times 1 = 9$
18. $54 \div 6 = 9$
$9 \times 5 = 45$
19. 24 hours $\div 6 = 4$ hours
4 hours $\times 1 = 4$ hours
20. 4×60 minutes $= 240$ minutes

Lesson Practice 2A

1. 10^3 ; 10^2 ; 10^1 ; 10^0
2. $10 \times 10 \times 10 \times 10 \times 10 \times 10 = 1,000,000$
3. 10
4. $10 \times 10 \times 10 \times 10 = 10,000$
5. $10 \times 10 \times 10 \times 10 \times 10 = 100,000$
6. 10^4
7. 10^6
8. 10^1

9. 10^2
10. done
11. $2 \times 100 + 7 \times 10 + 6 \times 1$;
    $2 \times 10^2 + 7 \times 10^1 + 6 \times 10^0$
12. $1 \times 1,000 + 4 \times 100 + 9 \times 1$;
    $1 \times 10^3 + 4 \times 10^2 + 9 \times 10^0$
13. $3 \times 10,000 + 1 \times 1,000 + 5 \times 100$;
    $3 \times 10^4 + 1 \times 10^3 + 5 \times 10^2$
14. 8,403
15. 70,060
16. 4,962
17. 3,530
18. 52,174

Lesson Practice 2B

1. 10^3 ; 10^2 ; 10^1 ; 10^0
2. $10 \times 10 = 100$
3. $10 \times 10 \times 10 \times 10 \times 10 = 100,000$
4. $10 \times 10 \times 10 = 1,000$
5. $10 \times 10 \times 10 \times 10 \times 10 \times 10 = 1,000,000$
6. 10^0
7. 10^2
8. 10^4
9. 10^3
10. $4 \times 1,000 + 8 \times 100 + 3 \times 10 + 6 \times 1$;
    $4 \times 10^3 + 8 \times 10^2 + 3 \times 10^1 + 6 \times 10^0$
11. $6 \times 100,000 + 2 \times 100 + 7 \times 10 + 5 \times 1$
    $6 \times 10^5 + 2 \times 10^2 + 7 \times 10^1 + 5 \times 10^0$
12. $3 \times 100 + 8 \times 10 + 4 \times 1$;
    $3 \times 10^2 + 8 \times 10^1 + 4 \times 10^0$
13. 5×10;
    5×10^1
14. 9,349
15. 617
16. 40,703
17. 2,874
18. 12,211

Lesson Practice 2C

1. 10^3 ; 10^2 ; 10^1 ; 10^0
2. $10 \times 10 \times 10 \times 10 = 10,000$
3. $10 \times 10 = 100$
4. $10 \times 10 \times 10 \times 10 \times 10 \times 10 = 1,000,000$
5. 1
6. 10^5
7. 10^1
8. 10^3
9. 10^6
10. $7 \times 100 + 2 \times 1$;
    $7 \times 10^2 + 2 \times 10^0$
11. $1 \times 10,000 + 1 \times 1,000 + 6 \times 100 + 8 \times 1$;
    $1 \times 10^4 + 1 \times 10^3 + 6 \times 10^2 + 8 \times 10^0$
12. $8 \times 1,000,000$;
    8×10^6
13. $4 \times 10 + 8 \times 1$;
    $4 \times 10^1 + 8 \times 10^0$
14. 5,607
15. 1,980
16. 770,000
17. 3,610,000
18. 216,534

Systematic Review 2D

1. 10,000
2. 100
3. 1,000
4. 0
5. 2
6. 4
7. $3 \times 1,000 + 7 \times 100 + 6 \times 10 + 6 \times 1$;
    $3 \times 10^3 + 7 \times 10^2 + 6 \times 10^1 + 6 \times 10^0$
8. $5 \times 10,000 + 1 \times 1,000 + 1 \times 10 + 7 \times 1$;
    $5 \times 10^4 + 1 \times 10^3 + 1 \times 10^1 + 7 \times 10^0$
9. 6,220
10. 10,901
11. $1^2 = 1$
12. $2^3 = 8$
13. $2^4 = 16$

14. $\dfrac{1}{3} = \dfrac{2}{6} = \dfrac{3}{9} = \dfrac{4}{12}$

15. $\dfrac{2}{5} = \dfrac{4}{10} = \dfrac{6}{15} = \dfrac{8}{20}$

16. $42 \div 7 = 6$;
$6 \times 5 = 30$ ducks

17. 12 months $\div 4 = 3$ months
The letter is J

18. $24 \div 6 = 4$;
$4 \times 5 = 20$ right-handed students
$24 - 20 = 4$ left-handed students

Systematic Review 2E

1. 1

2. 1,000

3. 10

4. 10^1

5. 10^3

6. 10^0

7. $7 \times 100 + 4 \times 10 + 8 \times 1$;
$7 \times 10^2 + 4 \times 10^1 + 8 \times 10^0$

8. $1 \times 10,000 + 2 \times 1,000 + 4 \times 100 + 6 \times 10 + 8 \times 1$;
$1 \times 10^4 + 2 \times 10^3 + 4 \times 10^2 + 6 \times 10^1 + 8 \times 10^0$

9. 8,437

10. 60,294

11. $7^2 = 49$

12. $54^1 = 54$

13. $3^3 = 27$

14. $\dfrac{5}{6} = \dfrac{10}{12} = \dfrac{15}{18} = \dfrac{20}{24}$ → *printed as 20 in some books*

15. $\dfrac{1}{4} = \dfrac{2}{8} = \dfrac{3}{12} = \dfrac{4}{16}$

16. $\dfrac{3}{8} = \dfrac{6}{16} = \dfrac{9}{24} = \dfrac{12}{32}$

17. $\dfrac{6}{7} = \dfrac{12}{14} = \dfrac{18}{21} = \dfrac{24}{28}$

18. 24 hours $\div 12 = 2$ hours;
2 hours $\times 5 = 10$ hours

19. 12 months $\div 4 = 3$ months;
3 months $\times 1 = 3$ months

20. 28 days $\div 4 = 7$ days;
7 days $\times 3 = 21$ days

Systematic Review 2F

1. 100,000

2. 10,000

3. 100

4. 10^3

5. 10^4

6. 10^2

7. $5 \times 1,000 + 8 \times 100 + 8 \times 10 + 9 \times 1$;
$5 \times 10^3 + 8 \times 10^2 + 8 \times 10^1 + 9 \times 10^0$

8. $6 \times 10,000 + 4 \times 100 + 1 \times 10$;
$6 \times 10^4 + 4 \times 10^2 + 1 \times 10^1$

9. 7,260

10. 55,007

11. $9^2 = 81$

12. $1^5 = 1$

13. $2^4 = 16$

14. $\dfrac{2}{3} = \dfrac{4}{6} = \dfrac{6}{9} = \dfrac{8}{12}$

15. $\dfrac{3}{5} = \dfrac{6}{10} = \dfrac{9}{15} = \dfrac{12}{20}$

16. $\dfrac{1}{9} = \dfrac{2}{18} = \dfrac{3}{27} = \dfrac{4}{36}$

17. $\dfrac{7}{10} = \dfrac{14}{20} = \dfrac{21}{30} = \dfrac{28}{40}$

18. 60 minutes $\div 3 = 20$ minutes;
20 minutes $\times 2 = 40$ minutes

19. $75 \div 5 = 15$;
$15 \times 4 = 60$ bean plants

20. $30 \div 10 = 3$ packages damaged
$30 - 3 = 27$ packages arrived safely

Lesson Practice 3A

1. 10; 1; $\dfrac{1}{100}$; $\dfrac{1}{1,000}$

2. multiply

3. divide

4. done

5. $6 \times 10 + 7 \times 1 + 2 \times \dfrac{1}{10} + 1 \times \dfrac{1}{100}$

6. done

7. $2 \times 10^0 + 7 \times \dfrac{1}{10^2}$

8. 1,643.119

9. 371.045
10. 68.95
11. done
12. dollar; dimes; cents; $1.00 + $0.30 + $0.08
13. dollar; dimes; cents; $1.00 + $0.70 + $0.02
14. dime
15. cent

Lesson Practice 3B

1. $1,000; 100; 1; \dfrac{1}{10}; \dfrac{1}{1,000}$
2. divide
3. multiply
4. $1 \times 1 + 9 \times \dfrac{1}{10} + 9 \times \dfrac{1}{100} + 9 \times \dfrac{1}{1,000}$
5. $2 \times 10 + 3 \times 1 + 6 \times \dfrac{1}{10} + 5 \times \dfrac{1}{100}$
6. $2 \times 10^2 + 3 \times 10^0 + 1 \times \dfrac{1}{10^2} + 6 \times \dfrac{1}{10^3}$
7. $8 \times 10^3 + 4 \times 10^2 + 3 \times 10^1 + 9 \times 10^0 + 7 \times \dfrac{1}{10^1}$
8. 2,401.613
9. 770.901
10. 5.008
11. dollars; dime; cents;
    $0.10; $0.09; $2.19
12. 3; 2; 7; $3.00 + $0.20 + $0.07 = $3.27
13. dimes; cents; $0.40 + $0.05 = $0.45
14. cent
15. dime

Lesson Practice 3C

1. $1,000; 100; 10; \dfrac{1}{10}; \dfrac{1}{100}$
2. left
3. right
4. $3 \times 100 + 4 \times 1 + 5 \times \dfrac{1}{100}$
5. $4 \times 1 + 6 \times \dfrac{1}{10} + 7 \times \dfrac{1}{100} + 9 \times \dfrac{1}{1,000}$
6. $6 \times 10^2 + 9 \times 10^1 + 1 \times 10^0 + 4 \times \dfrac{1}{10^1}$
7. $2 \times 10^1 + 5 \times 10^0 + 3 \times \dfrac{1}{10^1}$

8. 9,841.132
9. 3,006.084
10. 200.5
11. dollars; dimes; cents
    $7.00 + $0.40 + $0.05 = $7.45
12. 1; 1; 4; $1.00 + $0.10 + $0.04 = $1.14
13. dimes; cent; $0.60 + $0.01 = $0.61
14. dollar
15. dollar

Systematic Review 3D

1. $2 \times 10^3 + 3 \times 10^2 + 1 \times \dfrac{1}{10^1}$
2. $3 \times 10^1 + 8 \times 10^0 + 1 \times \dfrac{1}{10^1} + 2 \times \dfrac{1}{10^2} + 3 \times \dfrac{1}{10^3}$
3. 8,715.546
4. 6.411
5. dollars; dime; cents; $3.00; $0.09; $3.19
6. 9; 4; $9.00 + $0.04 = $9.04
7. $5^2 = 25$
8. $1^5 = 1$
9. $10^3 = 1,000$
10. $10^0 = 1$
11. $\dfrac{1}{2} = \dfrac{2}{4} = \dfrac{3}{6} = \dfrac{4}{8}$
12. $\dfrac{5}{8} = \dfrac{10}{16} = \dfrac{15}{24} = \dfrac{20}{32}$
13. $\dfrac{8}{10} \div \dfrac{2}{2} = \dfrac{4}{5}$
14. $\dfrac{4}{24} \div \dfrac{4}{4} = \dfrac{1}{6}$
15. $\dfrac{6}{18} \div \dfrac{6}{6} = \dfrac{1}{3}$
16. $\dfrac{18}{30} \div \dfrac{6}{6} = \dfrac{3}{5}$
17. $4.00 + $0.30 + $0.06 = $4.36
18. $100¢ \div 10 = 10¢$;
    $10¢ \times 4 = 40¢$

Systematic Review 3E

1. $6 \times \dfrac{1}{10^1} + 1 \times \dfrac{1}{10^2} + 5 \times \dfrac{1}{10^3}$
2. $1 \times 10^2 + 3 \times 10^1 + 5 \times 10^0 + 4 \times \dfrac{1}{10^3}$

3. 451.221

4. 10.607

5. dollar; dimes; cents;
$1.00 + $0.00 + $0.09 = $1.09

6. 6; 7; 5; $6.00 + $0.70 + $0.05 = $6.75

7. $3^2 = 9$

8. $4^3 = 64$

9. $2^2 = 4$

10. $10^3 = 1,000$

11. $\dfrac{4}{5} = \dfrac{8}{10} = \dfrac{12}{15} = \dfrac{16}{20}$

12. $\dfrac{3}{7} = \dfrac{6}{14} = \dfrac{9}{21} = \dfrac{12}{28}$

13. $\dfrac{11}{22} \div \dfrac{11}{11} = \dfrac{1}{2}$

14. $\dfrac{5}{25} \div \dfrac{5}{5} = \dfrac{1}{5}$

15. $\dfrac{4}{16} \div \dfrac{4}{4} = \dfrac{1}{4}$

16. $\dfrac{8}{32} \div \dfrac{8}{8} = \dfrac{1}{4}$

17. $1.00 + $0.40 + $0.07 = $1.47

18. 100¢ ÷ 5 = 20¢;
20¢ × 3 = 60¢

19. $\dfrac{2}{4} = \dfrac{1}{2}$ of a melon

20. $10^2 = 100$ blocks

Systematic Review 3F

1. $1 \times 10^0 + 1 \times \dfrac{1}{10^3}$

2. $1 \times 10^3 + 3 \times 10^2 + 5 \times 10^1 +$
$8 \times 10^0 + 9 \times \dfrac{1}{10^1} + 1 \times \dfrac{1}{10^2}$

3. 6,528.05

4. 2,000.986

5. dollars; dimes; cents;
$9.00 + $0.80 + $0.07 = $9.87

6. dollars; dimes; cents;
$2.00 + $0.00 + $0.08 = $2.08

7. $3^4 = 81$

8. $1^3 = 1$

9. $10^0 = 1$

10. $5^2 = 25$

11. $\dfrac{9}{10} = \dfrac{18}{20} = \dfrac{27}{30} = \dfrac{36}{40}$

12. $\dfrac{1}{6} = \dfrac{2}{12} = \dfrac{3}{18} = \dfrac{4}{24}$

13. $\dfrac{5}{30} \div \dfrac{5}{5} = \dfrac{1}{6}$

14. $\dfrac{14}{35} \div \dfrac{7}{7} = \dfrac{2}{5}$

15. $\dfrac{20}{40} \div \dfrac{20}{20} = \dfrac{1}{2}$

16. $\dfrac{18}{27} \div \dfrac{9}{9} = \dfrac{2}{3}$

17. $6.00 + $0.09 = $6.09

18. 100¢ ÷ 100 = 1¢;
1¢ × 7 = 7¢

19. $\dfrac{9}{12} = \dfrac{3}{4}$ of the dishes

20. 20 questions ÷ 5 = 4;
4 × 4 = 16 questions

Lesson Practice 4A

1. done

2. done

3. $\begin{array}{r} 1.53 \\ +1.12 \\ \hline 2.65 \end{array}$

4. $\begin{array}{r} 2.17 \\ + .31 \\ \hline 2.48 \end{array}$

5. $\begin{array}{r} 1.8 \\ +1.0 \\ \hline 2.8 \end{array}$

6. $\begin{array}{r} 3.2 \\ + .4 \\ \hline 3.6 \end{array}$

7. $\begin{array}{r} ^1 \\ 1.13 \\ +1.68 \\ \hline 2.81 \end{array}$

8. $\begin{array}{r} ^1 \\ 1.67 \\ + .42 \\ \hline 2.09 \end{array}$

9.   1.5
   +1.2
    2.7

10.   2.1
   + .8
    2.9

11.   1
   1.16
  +1.46
   2.62

12.   3.90
  + .02
   3.92

13.   1
   2.6
  +1.5
   4.1

14.   1
   1.8
  +1.3
   3.1

15.   3.00
  +1.62
   4.62

16.   4.48
  + .10
   4.58

17. $4.51
  +$ .35
   $4.86

18.   1
   1.5
  +2.72
  4.22 miles

Lesson Practice 4B

1.   1
   7.1
  + 6.2
   13.3

2.   1
   5.9
  +1.2
   7.1

3.   1
   2.45
  +5.07
   7.52

4.   1
   4.13
  +1.96
   6.09

5.   7.0
  +2.8
   9.8

6.   1.5
  + 9.3
  10.8

7.   1
   8.84
  + 3.09
  11.93

8.   .437
  +.250
  .687

9.   1
   8.8
  + 3.4
  12.2

10.   6.2
  + .4
   6.6

11.   1
   2.70
  + 9.41
  12.11

12.   1
   5.52
  + .60
   6.12

13.   3.9
  +4.0
   7.9

14.   1
   7.5
  + .8
   8.3

15.   4.15
  +3.00
   7.15

16.
$$
\begin{array}{r}
^{1\ 1}\\
.524\\
+.277\\
\hline
.801
\end{array}
$$

17.
$$
\begin{array}{r}
^{1}\\
\$12.95\\
+\$15.50\\
\hline
\$28.45
\end{array}
$$

18.
$$
\begin{array}{r}
^{1}\\
.625\\
+2.125\\
\hline
2.750 \text{ gallons}
\end{array}
$$

Lesson Practice 4C

1.
$$
\begin{array}{r}
^{1}\\
3.0\\
+\ 9.8\\
\hline
12.8
\end{array}
$$

2.
$$
\begin{array}{r}
7.1\\
+1.3\\
\hline
8.4
\end{array}
$$

3.
$$
\begin{array}{r}
^{1\ 1}\\
1.95\\
+\ 8.15\\
\hline
10.10
\end{array}
$$

4.
$$
\begin{array}{r}
^{1}\\
3.51\\
+2.68\\
\hline
6.19
\end{array}
$$

5.
$$
\begin{array}{r}
^{1}\\
5.9\\
+\ .4\\
\hline
6.3
\end{array}
$$

6.
$$
\begin{array}{r}
4.1\\
+3.0\\
\hline
7.1
\end{array}
$$

7.
$$
\begin{array}{r}
^{1}\\
2.34\\
+\ .71\\
\hline
3.05
\end{array}
$$

8.
$$
\begin{array}{r}
.440\\
+.300\\
\hline
.740
\end{array}
$$

9.
$$
\begin{array}{r}
^{1}\\
6.5\\
+\ 5.0\\
\hline
11.5
\end{array}
$$

10.
$$
\begin{array}{r}
^{1}\\
2.8\\
+5.9\\
\hline
8.7
\end{array}
$$

11.
$$
\begin{array}{r}
^{1\ 1}\\
7.48\\
+1.93\\
\hline
9.41
\end{array}
$$

12.
$$
\begin{array}{r}
.162\\
+8.000\\
\hline
8.162
\end{array}
$$

13.
$$
\begin{array}{r}
8.7\\
+\ 8.1\\
\hline
16.8
\end{array}
$$

14.
$$
\begin{array}{r}
6.0\\
+\ .1\\
\hline
6.1
\end{array}
$$

15.
$$
\begin{array}{r}
^{1}\\
.731\\
+\ .402\\
\hline
1.133
\end{array}
$$

16.
$$
\begin{array}{r}
1.125\\
+\ .112\\
\hline
1.237
\end{array}
$$

17.
$$
\begin{array}{r}
4.3\\
+\ .5\\
\hline
4.8 \text{ bushels}
\end{array}
$$

18.
$$
\begin{array}{r}
2.045\\
+\ .5\\
\hline
2.545 \text{ inches}
\end{array}
$$

Systematic Review 4D

1.
$$
\begin{array}{r}
^{1}\\
1.5\\
+\ 9.3\\
\hline
10.8
\end{array}
$$

2.
$$
\begin{array}{r}
^{1}\\
5.9\\
+1.6\\
\hline
7.5
\end{array}
$$

3.
$$
\begin{array}{r}
6.34\\
+2.41\\
\hline
8.75
\end{array}
$$

4.
$$\overset{1}{}1.82$$
$$+\ 9.3$$
$$\overline{11.12}$$

5. $2^3 = 8$

6. $6^2 = 36$

7. $10^4 = 10,000$

8. $7^2 = 49$

9. $1 \times 100 + 7 \times 10 + 6 \times 1 + 2 \times \dfrac{1}{10} + 1 \times \dfrac{1}{100}$

10. $6 \times \dfrac{1}{10} + 8 \times \dfrac{1}{100} + 5 \times \dfrac{1}{1,000}$

11. $4 \times 1 + 5 \times \dfrac{1}{10}$

12. $\dfrac{1}{4} = \dfrac{2}{8} = \dfrac{3}{12} = \dfrac{4}{16}$

13. $\dfrac{5}{8} = \dfrac{10}{16} = \dfrac{15}{24} = \dfrac{20}{32}$

14. $\dfrac{1}{4} + \dfrac{3}{5} = \dfrac{5}{20} + \dfrac{12}{20} = \dfrac{17}{20}$

15. $\dfrac{3}{4} + \dfrac{1}{6} = \dfrac{18}{24} + \dfrac{4}{24} = \dfrac{22}{24} = \dfrac{11}{12}$

16. $\dfrac{1}{3} + \dfrac{2}{5} = \dfrac{5}{15} + \dfrac{6}{15} = \dfrac{11}{15}$

17.
$$\overset{1\ 1}{}75.25$$
$$+\ 1.75$$
$$\overline{77.00 \text{ inches}}$$

18. $12 \div 6 = 2$ spoiled apples
$12 - 2 = 10$ good apples

Systematic Review 4E

1.
$$\overset{1\ 1}{}8.6$$
$$+\ 2.4$$
$$\overline{11.0}$$

2.
$$3.0$$
$$+4.4$$
$$\overline{7.4}$$

3.
$$\overset{1\ \ \ 1}{}3.07$$
$$+\ 9.25$$
$$\overline{12.32}$$

4.
$$5.00$$
$$+3.24$$
$$\overline{8.24}$$

5. $3^4 = 81$

6. $5^2 = 25$

7. $1^7 = 1$

8. $10^3 = 1,000$

9. $4 \times 10^1 + 3 \times 10^0 + 3 \times \dfrac{1}{10^1}$

10. $6 \times 10^0 + 1 \times \dfrac{1}{10^1} + 5 \times \dfrac{1}{10^3}$

11. $2 \times 10^2 + 3 \times \dfrac{1}{10^1} + 4 \times \dfrac{1}{10^2}$

12. $\dfrac{1}{2} = \dfrac{2}{4} = \dfrac{3}{6} = \dfrac{4}{8}$

13. $\dfrac{9}{10} = \dfrac{18}{20} = \dfrac{27}{30} = \dfrac{36}{40}$

14. $\dfrac{1}{9} + \dfrac{1}{2} = \dfrac{2}{18} + \dfrac{9}{18} = \dfrac{11}{18}$

15. $\dfrac{2}{5} + \dfrac{5}{6} = \dfrac{12}{30} + \dfrac{25}{30} = \dfrac{37}{30} = 1\dfrac{7}{30}$

16. $\dfrac{1}{10} + \dfrac{2}{3} = \dfrac{3}{30} + \dfrac{20}{30} = \dfrac{23}{30}$

17.
$$.5$$
$$+.25$$
$$\overline{.75 \text{ hours}}$$

18.
$$\overset{1\ 1}{}9.5$$
$$+11.6$$
$$\overline{21.1 \text{ gallons}}$$

19. $\dfrac{2}{3} + \dfrac{1}{5} = \dfrac{10}{15} + \dfrac{3}{15} = \dfrac{13}{15}$ of the problems

20. $30 \div 15 = 2$;
$2 \times 13 = 26$ problems

Systematic Review 4F

1.
$$5.6$$
$$+4.3$$
$$\overline{9.9}$$

2.
$$\overset{1\ 1}{}1.9$$
$$+\ 9.2$$
$$\overline{11.1}$$

3.
$$\overset{1}{}5.13$$
$$+\ 9.50$$
$$\overline{14.63}$$

4.
$$\overset{1\ 1}{}4.17$$
$$+1.95$$
$$\overline{6.12}$$

5. $8^2 = 64$

6. $10^0 = 1$

7. $4^3 = 64$

8. $9^2 = 81$

9. $9,500.1$

10. 158.004

11. $\dfrac{1}{3} = \dfrac{2}{6} = \dfrac{3}{9} = \dfrac{4}{12}$

12. $\dfrac{3}{7} = \dfrac{6}{14} = \dfrac{9}{21} = \dfrac{12}{28}$

13. $\dfrac{2}{7} + \dfrac{1}{8} = \dfrac{16}{56} + \dfrac{7}{56} = \dfrac{23}{56}$

14. $\dfrac{3}{5} + \dfrac{2}{9} = \dfrac{27}{45} + \dfrac{10}{45} = \dfrac{37}{45}$

15. $\dfrac{3}{4} + \dfrac{1}{5} = \dfrac{15}{20} + \dfrac{4}{20} = \dfrac{19}{20}$

16.
$$\begin{array}{r} \overset{1}{}\$2.25 \\ +\$1.69 \\ \hline \$3.94 \end{array}$$

17.
$$\begin{array}{r} \overset{1}{}\$\ 4.00 \\ \$\ 2.50 \\ +\$\ 8.35 \\ \hline \$14.85 \end{array}$$

18. $\dfrac{5}{15} = \dfrac{1}{3}$

19. $\dfrac{3}{8} + \dfrac{1}{3} = \dfrac{9}{24} + \dfrac{8}{24} = \dfrac{17}{24}$; no

20. $27 \div 9 = 3;$
    $3 \times 5 = 15$ players

Lesson Practice 5A

1. done

2.
$$\begin{array}{r} 1\overset{7}{\cancel{8}}.7 \\ -\ \ 4.8 \\ \hline 13.9 \end{array}$$

3.
$$\begin{array}{r} \overset{8}{\cancel{9}}.\overset{12}{\cancel{3}}5 \\ -\ 8.46 \\ \hline .89 \end{array}$$

4.
$$\begin{array}{r} \overset{1}{\cancel{2}}\overset{11}{\cancel{2}}.0 \\ -\ \ 9.6 \\ \hline 12.4 \end{array}$$

5.
$$\begin{array}{r} 6.4 \\ -5.3 \\ \hline 1.1 \end{array}$$

6.
$$\begin{array}{r} \overset{4}{\cancel{5}}.0 \\ -2.4 \\ \hline 2.6 \end{array}$$

7.
$$\begin{array}{r} \overset{1}{2}3.\overset{5}{\cancel{6}}0 \\ -\ \ 9.43 \\ \hline 14.17 \end{array}$$

8.
$$\begin{array}{r} 2.8\overset{7}{\cancel{1}} \\ -\ .63 \\ \hline 2.18 \end{array}$$

9.
$$\begin{array}{r} \overset{3}{\cancel{4}}.0 \\ -2.6 \\ \hline 1.4 \end{array}$$

10.
$$\begin{array}{r} \overset{8}{\cancel{9}}.6 \\ -\ .9 \\ \hline 8.7 \end{array}$$

11.
$$\begin{array}{r} 8.93 \\ -5.00 \\ \hline 3.93 \end{array}$$

12.
$$\begin{array}{r} \overset{3}{\cancel{4}}.\overset{16}{\cancel{7}}0 \\ -1.98 \\ \hline 2.72 \end{array}$$

13.
$$\begin{array}{r} \$\overset{6}{\cancel{7}}.\overset{10}{\cancel{1}}7 \\ -\$2.98 \\ \hline \$4.19 \end{array}$$

14.
$$\begin{array}{r} 5.75 \\ -1.50 \\ \hline 4.25 \text{ feet} \end{array}$$

15.
$$\begin{array}{r} 15\overset{4}{\cancel{0}}.\overset{9}{\cancel{0}} \\ -\ \ 4.5 \\ \hline 145.5 \text{ pounds} \end{array}$$

16.
$$\begin{array}{r} 15.\overset{4}{\cancel{1}}\overset{10}{\cancel{5}} \\ -\ 4.29 \\ \hline 10.86 \text{ seconds} \end{array}$$

17.
$$\begin{array}{r} 9 \\ 1\overset{9}{\cancel{0}}.{}^{1}2 \\ -.7 \\ \hline 9.5 \text{ miles} \end{array}$$

18.
$$\begin{array}{r} {}^{9}{}^{14} \\ \$1\overset{9}{\cancel{0}}.\overset{14}{\cancel{5}}{}^{1}0 \\ -\$4.79 \\ \hline \$5.71 \end{array}$$

Lesson Practice 5B

1.
$$\begin{array}{r} 8.3 \\ -.3 \\ \hline 8.0 \end{array}$$

2.
$$\begin{array}{r} 8 \\ \overset{8}{\cancel{9}}.{}^{1}1 \\ -6.3 \\ \hline 2.8 \end{array}$$

3.
$$\begin{array}{r} 49 \\ \overset{4}{\cancel{5}}.\overset{9}{\cancel{0}}{}^{1}0 \\ -2.33 \\ \hline 2.67 \end{array}$$

4.
$$\begin{array}{r} 6.98 \\ -1.90 \\ \hline 5.08 \end{array}$$

5.
$$\begin{array}{r} 4 \\ \overset{4}{\cancel{5}}.{}^{1}1 \\ -4.8 \\ \hline .3 \end{array}$$

6.
$$\begin{array}{r} 9.2 \\ -1.0 \\ \hline 8.2 \end{array}$$

7.
$$\begin{array}{r} 5 \\ 8.\overset{5}{\cancel{6}}{}^{1}0 \\ -5.35 \\ \hline 3.25 \end{array}$$

8.
$$\begin{array}{r} 6 \\ 12.\overset{6}{\cancel{7}}{}^{1}5 \\ -9.06 \\ \hline 3.69 \end{array}$$

9.
$$\begin{array}{r} 1.3 \\ -.6 \\ \hline .7 \end{array}$$

10.
$$\begin{array}{r} 5 \\ \overset{5}{\cancel{6}}.{}^{1}2 \\ -1.8 \\ \hline 4.4 \end{array}$$

11.
$$\begin{array}{r} 410 \\ .\overset{4}{\cancel{5}}\overset{10}{\cancel{1}}{}^{1}3 \\ -.026 \\ \hline .487 \end{array}$$

12.
$$\begin{array}{r} .362 \\ -.100 \\ \hline .262 \end{array}$$

13.
$$\begin{array}{r} 0109 \\ \$1\overset{0}{\cancel{1}}.\overset{10}{\cancel{0}}\overset{9}{\cancel{0}}{}^{1}0 \\ -\$5.75 \\ \hline \$5.25 \end{array}$$

14.
$$\begin{array}{r} 4 \\ \$3.\overset{4}{\cancel{5}}{}^{1}0 \\ -\$2.45 \\ \hline \$1.05 \end{array}$$

15.
$$\begin{array}{r} 7 \\ \overset{7}{\cancel{8}}.{}^{1}5 \\ -7.6 \\ \hline .9 \text{ hours} \end{array}$$

16.
$$\begin{array}{r} 10 \\ 1\overset{10}{\cancel{1}}.{}^{1}0 \\ -4.5 \\ \hline 6.5 \text{ miles} \end{array}$$

17.
$$\begin{array}{r} 39 \\ 12.\overset{3}{\cancel{4}}\overset{9}{\cancel{0}}{}^{1}0 \\ -2.375 \\ \hline 10.025 \text{ years} \end{array}$$

18.
$$\begin{array}{r} 110 \\ \overset{1}{\cancel{2}}.\overset{10}{\cancel{1}}{}^{1}0 \\ -.16 \\ \hline 1.94 \text{ inches} \end{array}$$

Lesson Practice 5C

1.
$$\begin{array}{r} 7.8 \\ -1.2 \\ \hline 6.6 \end{array}$$

2.
$$
\begin{array}{r}
\overset{3}{\cancel{4}.^{1}1} \\
-\ 2.9 \\
\hline
1.2
\end{array}
$$

3.
$$
\begin{array}{r}
\overset{9}{\cancel{1}}0.\overset{9}{\cancel{0}}{}^{1}0 \\
-\ 2.6\ 7 \\
\hline
7.3\ 3
\end{array}
$$

4.
$$
\begin{array}{r}
\overset{4}{\cancel{5}}.\overset{9}{\cancel{0}}{}^{1}1 \\
-\ 1.9\ 2 \\
\hline
3.0\ 9
\end{array}
$$

5.
$$
\begin{array}{r}
\overset{5}{\cancel{6}}.{}^{1}0 \\
-\ .3 \\
\hline
5.7
\end{array}
$$

6.
$$
\begin{array}{r}
\overset{2}{\cancel{3}}.{}^{1}7 \\
-\ 2.8 \\
\hline
.9
\end{array}
$$

7.
$$
\begin{array}{r}
\overset{0}{1}\overset{11}{\cancel{1}}.\overset{}{\cancel{2}}{}^{1}0 \\
-\ 8.2\ 3 \\
\hline
2.9\ 7
\end{array}
$$

8.
$$
\begin{array}{r}
\overset{0}{\cancel{1}}.\overset{9}{\cancel{0}}{}^{1}0 \\
-\ .7\ 7 \\
\hline
.2\ 3
\end{array}
$$

9.
$$
\begin{array}{r}
\overset{8}{\cancel{9}}.{}^{1}2 \\
-\ .5 \\
\hline
8.7
\end{array}
$$

10.
$$
\begin{array}{r}
8.1 \\
-7.0 \\
\hline
1.1
\end{array}
$$

11.
$$
\begin{array}{r}
.706 \\
-.300 \\
\hline
.406
\end{array}
$$

12.
$$
\begin{array}{r}
.2\overset{8}{\cancel{9}}{}^{1}8 \\
-.0\ 0\ 9 \\
\hline
.28\ 9
\end{array}
$$

13.
$$
\begin{array}{r}
\overset{1}{\$}\overset{4}{\cancel{2}}\overset{9}{\cancel{5}}.\cancel{0}{}^{1}0 \\
-\$\ \ 94.7\ 6 \\
\hline
\$1\ 30.2\ 4
\end{array}
$$

14.
$$
\begin{array}{r}
\$\overset{9}{1}0.\overset{9}{\cancel{0}}{}^{1}0 \\
-\$\ 7.4\ 5 \\
\hline
\$\ 2.5\ 5
\end{array}
$$

15.
$$
\begin{array}{r}
6.7 \\
-4.3 \\
\hline
2.4\ \ \text{loaves}
\end{array}
$$

16.
$$
\begin{array}{r}
\overset{1}{\$}\overset{9}{\cancel{2}}\overset{9}{\cancel{0}}.\cancel{0}{}^{1}0 \\
-\$1\ \ 6.2\ 5 \\
\hline
\$\ \ 3.7\ 5
\end{array}
$$

17.
$$
\begin{array}{r}
\overset{2}{\cancel{3}}\overset{10}{\cancel{1}}.{}^{1}0 \\
-\ \ 7.5 \\
\hline
2\ 3.5\ \ \text{chapters}
\end{array}
$$

18.
$$
\begin{array}{r}
\overset{3}{\cancel{4}}{}^{1}5.6 \\
-\ \ 6.5 \\
\hline
3\ 9.1\ \ \text{ft}
\end{array}
$$

Systematic Review 5D

1.
$$
\begin{array}{r}
\overset{3}{\cancel{4}}.{}^{1}2 \\
-3.9 \\
\hline
.3
\end{array}
$$

2.
$$
\begin{array}{r}
8.\overset{5}{\cancel{6}}{}^{1}0 \\
-\ .0\ 4 \\
\hline
8.5\ 6
\end{array}
$$

3.
$$
\begin{array}{r}
.007 \\
-.002 \\
\hline
.005
\end{array}
$$

4.
$$
\begin{array}{r}
{}^{1}99 \\
+\ .02 \\
\hline
1.01
\end{array}
$$

5.
$$
\begin{array}{r}
{}^{1}5.7 \\
+2.3 \\
\hline
8.0
\end{array}
$$

6.
$$6.025$$
$$+\ .800$$
$$6.825$$

7. $\dfrac{1}{6} + \dfrac{4}{6} = \dfrac{5}{6}$

8. $\dfrac{2}{9} + \dfrac{1}{10} = \dfrac{20}{90} + \dfrac{9}{90} = \dfrac{29}{90}$

9. $\dfrac{2}{5} + \dfrac{3}{8} = \dfrac{16}{40} + \dfrac{15}{40} = \dfrac{31}{40}$

10. $1 \times 1,000 + 3 \times 100 + 4 \times \dfrac{1}{10}$

11. $1 \times 1 + 9 \times \dfrac{1}{10} + 7 \times \dfrac{1}{100} + 8 \times \dfrac{1}{1,000}$

12. done

13. $\dfrac{2}{3} - \dfrac{1}{4} = \dfrac{8}{12} - \dfrac{3}{12} = \dfrac{5}{12}$

14. $\dfrac{5}{6} - \dfrac{3}{7} = \dfrac{35}{42} - \dfrac{18}{42} = \dfrac{17}{42}$

15.
$$\overset{9}{1\cancel{0}.}^{1}25$$
$$-\ 9\ .\ 75$$
$$.\ 50\ \text{mi}$$

16.
$$\$15.00 \qquad \$3\overset{5}{\cancel{6}}.\overset{16}{\cancel{7}}5$$
$$+\$21.75 \qquad -\$3\ 1\ .\ 9\ 9$$
$$\$36.75 \qquad\ \ \$\ \ 4\ .\ 7\ 6$$

17. $\dfrac{2}{3} - \dfrac{1}{6} = \dfrac{12}{18} - \dfrac{3}{18} = \dfrac{9}{18} = \dfrac{1}{2}$ of a pie

18. $\dfrac{1}{2} = \dfrac{2}{4}$ they got the same amount

Systematic Review 5E

1.
$$\overset{7}{\cancel{8}}.^{1}16$$
$$-\ 2\ .70$$
$$5\ .46$$

2.
$$\overset{8}{\cancel{9}}.^{1}04$$
$$-\ 2\ .\ 90$$
$$6\ .\ 14$$

3.
$$3.6\overset{3}{\cancel{4}}\,^{1}3$$
$$-2.0\ 0\ 8$$
$$1.6\ 3\ 5$$

4.
$$\overset{1}{\ }$$
$$6.9$$
$$+1.2$$
$$8.1$$

5.
$$4.007$$
$$+\ .932$$
$$4.939$$

6.
$$\overset{1}{\ }$$
$$31.600$$
$$+\ \ .456$$
$$32.056$$

7. $\dfrac{1}{8} + \dfrac{7}{9} = \dfrac{9}{72} + \dfrac{56}{72} = \dfrac{65}{72}$

8. $\dfrac{4}{5} + \dfrac{2}{11} = \dfrac{44}{55} + \dfrac{10}{55} = \dfrac{54}{55}$

9. $\dfrac{1}{3} + \dfrac{1}{4} = \dfrac{4}{12} + \dfrac{3}{12} = \dfrac{7}{12}$

10. $2 \times 10^1 + 8 \times 10^0 + 7 \times \dfrac{1}{10^1} + 8 \times \dfrac{1}{10^2}$

11. 2×10^4

12. $\dfrac{8}{9} - \dfrac{5}{9} = \dfrac{3}{9} = \dfrac{1}{3}$

13. $\dfrac{4}{5} - \dfrac{1}{2} = \dfrac{8}{10} - \dfrac{5}{10} = \dfrac{3}{10}$

14. $\dfrac{7}{10} - \dfrac{1}{6} = \dfrac{42}{60} - \dfrac{10}{60} = \dfrac{32}{60} = \dfrac{8}{15}$

15.
$$7.\overset{4}{\cancel{5}}\,^{1}0$$
$$-5.2\ 5$$
$$2.2\ 5\ \text{cups}$$

16.
$$\overset{1}{\ }$$
$$.90$$
$$1.25$$
$$.30$$
$$+\ \ .50$$
$$2.95\ \text{miles}$$

17.
$$4.5$$
$$+13.2$$
$$17.7\ \text{inches cut off}$$

$$\overset{1}{\cancel{2}}\overset{13}{\cancel{4}}.^{1}0$$
$$-1\ 7\ .7$$
$$6\ .\ 3\ \text{inches left}$$

18. $\dfrac{10}{100} = \dfrac{1}{10}$

19. $100¢ \div 10 = 10¢;$
$10¢ \times 5 = 50¢$

20. $\frac{2}{5} + \frac{1}{2} = \frac{4}{10} + \frac{5}{10} = \frac{9}{10}$ are gone

$1 - \frac{9}{10} = \frac{10}{10} - \frac{9}{10} = \frac{1}{10}$ on the job

Systematic Review 5F

1.
$$\overset{7}{\cancel{8}}.^{1}30$$
$$-\ 1.9$$
$$\overline{\quad 6.4}$$

2.
$$\overset{5}{\cancel{6}}.^{1}8$$
$$-\ 3.21$$
$$\overline{\quad 2.97}$$

3.
$$7.\overset{0}{\cancel{1}}\overset{11}{\cancel{2}}{}^{1}3$$
$$-4.0\ 4\ 5$$
$$\overline{\quad 3.0\ 7\ 8}$$

4.
$$\overset{1}{}$$
$$6.1$$
$$+\ \ .9$$
$$\overline{\quad 7.0}$$

5.
$$\overset{1}{}$$
$$3.930$$
$$+\ .605$$
$$\overline{\quad 4.535}$$

6.
$$28.700$$
$$+\ \ \ .008$$
$$\overline{\ 28.708}$$

7. $\frac{2}{4} + \frac{2}{5} = \frac{10}{20} + \frac{8}{20} = \frac{18}{20} = \frac{9}{10}$

8. $\frac{2}{3} + \frac{1}{4} = \frac{8}{12} + \frac{3}{12} = \frac{11}{12}$

9. $\frac{1}{6} + \frac{1}{5} = \frac{5}{30} + \frac{6}{30} = \frac{11}{30}$

10. 150,000.04

11. 6,800.22

12. $\frac{5}{8} - \frac{1}{3} = \frac{15}{24} - \frac{8}{24} = \frac{7}{24}$

13. $\frac{9}{10} - \frac{2}{5} = \frac{45}{50} - \frac{20}{50} = \frac{25}{50} = \frac{1}{2}$

14. $\frac{1}{4} - \frac{1}{7} = \frac{7}{28} - \frac{4}{28} = \frac{3}{28}$

15.
$$\overset{5}{\cancel{6}}.\overset{3}{\cancel{3}}\overset{}{\cancel{4}}{}^{1}5$$
$$-\ 4.73\ 8$$
$$\overline{\quad 1.\ 6\ 0\ 7}\ \text{ounces}$$

16.
$$\overset{1\ \ 1\ \ 1}{}$$
$$\$25.56$$
$$+\$\ \ 6.78$$
$$\overline{\ \$32.34}$$

17.
$$\$\overset{2}{\cancel{3}}\overset{}{2}.\overset{2}{\cancel{3}}{}^{1}4$$
$$-\$\ 1\ 6.1\ 6$$
$$\overline{\ \$\ 1\ 6.1\ 8}$$

18. $100¢ \div 100 = 1¢;$
$1¢ \times 17 = 17¢$

19. $\frac{1}{3} + \frac{1}{4} = \frac{4}{12} + \frac{3}{12} = \frac{7}{12}$ of the candy

20. $36 \div 12 = 3;$
$3 \times 7 = 21$ pieces given away
$36 - 21 = 15$ pieces left

Lesson Practice 6A

1. b: meter

2. c: liter

3. a: gram

4. f: 10

5. d: 100

6. e: 1,000

7. $\frac{1\ \text{kilogram(kg)}}{1,000\ \text{grams (g)}}$; $\frac{1\ \text{hectogram(hg)}}{100\ \text{grams (g)}}$;
$\frac{1\ \text{dekagram(dag)}}{10\ \text{grams (g)}}$; $\frac{1\ \text{gram(g)}}{1\ \text{gram(g)}}$

8. $\frac{1\ \text{kiloliter(kl)}}{1,000\ \text{liters (L)}}$; $\frac{1\ \text{hectoliter(hl)}}{100\ \text{liters (L)}}$;
$\frac{1\ \text{dekaliter(dal)}}{10\ \text{liters (L)}}$; $\frac{1\ \text{liter(L)}}{1\ \text{liter(L)}}$

9. $\frac{1\ \text{kilometer(km)}}{1,000\ \text{meters (m)}}$; $\frac{1\ \text{hectometer(hm)}}{100\ \text{meters (m)}}$;
$\frac{1\ \text{dekameter(dam)}}{10\ \text{meters (m)}}$; $\frac{1\ \text{meter(m)}}{1\ \text{meter(m)}}$

10. 1,000 meters

11. 100 liters

12. 10 grams

Lesson Practice 6B

1. meter
2. liter
3. gram
4. 1,000
5. 10
6. 100
7. $\dfrac{1 \text{ kilogram(kg)}}{1,000 \text{ grams (g)}}$; $\dfrac{1 \text{ hectogram(hg)}}{100 \text{ grams (g)}}$; $\dfrac{1 \text{ dekagram(dag)}}{10 \text{ grams (g)}}$; $\dfrac{1 \text{ gram(g)}}{1 \text{ gram(g)}}$
8. $\dfrac{1 \text{ kiloliter(kl)}}{1,000 \text{ liters (L)}}$; $\dfrac{1 \text{ hectoliter(hl)}}{100 \text{ liters (L)}}$; $\dfrac{1 \text{ dekaliter(dal)}}{10 \text{ liters (L)}}$; $\dfrac{1 \text{ liter(L)}}{1 \text{ liter(L)}}$
9. $\dfrac{1 \text{ kilometer(km)}}{1,000 \text{ meters (m)}}$; $\dfrac{1 \text{ hectometer(hm)}}{100 \text{ meters (m)}}$; $\dfrac{1 \text{ dekameter(dam)}}{10 \text{ meters (m)}}$; $\dfrac{1 \text{ meter(m)}}{1 \text{ meter(m)}}$
10. 100 grams
11. 1,000 liters
12. 10 meters

Lesson Practice 6C

1. c: meter
2. a: liter
3. b: gram
4. hecto
5. deka
6. kilo
7. $\dfrac{1 \text{ kilogram(kg)}}{1,000 \text{ grams (g)}}$; $\dfrac{1 \text{ hectogram(hg)}}{100 \text{ grams (g)}}$; $\dfrac{1 \text{ dekagram(dag)}}{10 \text{ grams (g)}}$; $\dfrac{1 \text{ gram(g)}}{1 \text{ gram(g)}}$
8. $\dfrac{1 \text{ kiloliter(kl)}}{1,000 \text{ liters (L)}}$; $\dfrac{1 \text{ hectoliter(hl)}}{100 \text{ liters (L)}}$; $\dfrac{1 \text{ dekaliter(dal)}}{10 \text{ liters (L)}}$; $\dfrac{1 \text{ liter(L)}}{1 \text{ liter(L)}}$
9. $\dfrac{1 \text{ kilometer(km)}}{1,000 \text{ meters (m)}}$; $\dfrac{1 \text{ hectometer(hm)}}{100 \text{ meters (m)}}$; $\dfrac{1 \text{ dekameter(dam)}}{10 \text{ meters (m)}}$; $\dfrac{1 \text{ meter(m)}}{1 \text{ meter(m)}}$
10. 10 liters
11. 100 meters
12. 1,000 grams

Systematic Review 6D

1. $\dfrac{1 \text{ kilogram(kg)}}{1,000 \text{ grams (g)}}$; $\dfrac{1 \text{ hectogram(hg)}}{100 \text{ grams (g)}}$; $\dfrac{1 \text{ dekagram(dag)}}{10 \text{ grams (g)}}$; $\dfrac{1 \text{ gram(g)}}{1 \text{ gram(g)}}$
2. $\dfrac{1 \text{ kiloliter(kl)}}{1,000 \text{ liters (L)}}$; $\dfrac{1 \text{ hectoliter(hl)}}{100 \text{ liters (L)}}$; $\dfrac{1 \text{ dekaliter(dal)}}{10 \text{ liters (L)}}$; $\dfrac{1 \text{ liter(L)}}{1 \text{ liter(L)}}$
3. $\dfrac{1 \text{ kilometer(km)}}{1,000 \text{ meters (m)}}$; $\dfrac{1 \text{ hectometer(hm)}}{100 \text{ meters (m)}}$; $\dfrac{1 \text{ dekameter(dam)}}{10 \text{ meters (m)}}$; $\dfrac{1 \text{ meter(m)}}{1 \text{ meter(m)}}$
4. $\begin{array}{r} 1 \\ 1.8 \\ +4.5 \\ \hline 6.3 \end{array}$
5. $\begin{array}{r} 6.4 \\ -5.3 \\ \hline 1.1 \end{array}$
6. $\begin{array}{r} 1 \\ 7.23 \\ +\ 3.54 \\ \hline 10.77 \end{array}$
7. $\begin{array}{r} 8.15 \\ -\ .13 \\ \hline 8.02 \end{array}$
8. $\dfrac{1}{6} + \dfrac{2}{3} = \dfrac{3}{18} + \dfrac{12}{18} = \dfrac{15}{18} = \dfrac{5}{6}$
9. $\dfrac{2}{5} + \dfrac{1}{2} = \dfrac{4}{10} + \dfrac{5}{10} = \dfrac{9}{10}$
10. $\dfrac{2}{3} - \dfrac{1}{4} = \dfrac{8}{12} - \dfrac{3}{12} = \dfrac{5}{12}$
11. $\dfrac{3}{5} - \dfrac{1}{3} = \dfrac{9}{15} - \dfrac{5}{15} = \dfrac{4}{15}$
12. $1\dfrac{5}{8} = \dfrac{13}{8}$
13. $4\dfrac{1}{2} = \dfrac{9}{2}$
14. $2\dfrac{3}{7} = \dfrac{17}{7}$
15. $5\dfrac{1}{3} = \dfrac{16}{3}$
16. $\begin{array}{r} \$5.95 \\ -\$1.50 \\ \hline \$4.45 \end{array}$
17. $1 \text{ km} = 1,000 \text{ m}$

18. $\dfrac{7}{8} - \dfrac{1}{4} = \dfrac{28}{32} - \dfrac{8}{32} =$

$\dfrac{20}{32} = \dfrac{5}{8}$ of a pizza

Systematic Review 6E

1. $\dfrac{1 \text{ kilogram(kg)}}{1,000 \text{ grams (g)}}$; $\dfrac{1 \text{ hectogram(hg)}}{100 \text{ grams (g)}}$;

$\dfrac{1 \text{ dekagram(dag)}}{10 \text{ grams (g)}}$; $\dfrac{1 \text{ gram(g)}}{1 \text{ gram(g)}}$

2. $\dfrac{1 \text{ kiloliter(kl)}}{1,000 \text{ liters (L)}}$; $\dfrac{1 \text{ hectoliter(hl)}}{100 \text{ liters (L)}}$;

$\dfrac{1 \text{ dekaliter(dal)}}{10 \text{ liters (L)}}$; $\dfrac{1 \text{ liter(L)}}{1 \text{ liter(L)}}$

3. $\dfrac{1 \text{ kilometer(km)}}{1,000 \text{ meters (m)}}$; $\dfrac{1 \text{ hectometer(hm)}}{100 \text{ meters (m)}}$;

$\dfrac{1 \text{ dekameter(dam)}}{10 \text{ meters (m)}}$; $\dfrac{1 \text{ meter(m)}}{1 \text{ meter(m)}}$

4.
$$
\begin{array}{r}
9.4 \\
+ \ .5 \\
\hline
9.9
\end{array}
$$

5.
$$
\begin{array}{r}
^6 \\
\not{7}.{}^1 0 \\
-3.1 \\
\hline
3.9
\end{array}
$$

6.
$$
\begin{array}{r}
{}^{1\ 1} \\
68.910 \\
+\ 2.306 \\
\hline
71.216
\end{array}
$$

7.
$$
\begin{array}{r}
{}^4 \\
.4\not{5}{}^1 3 \\
-.127 \\
\hline
.326
\end{array}
$$

8. $\dfrac{2}{4} + \dfrac{1}{3} = \dfrac{6}{12} + \dfrac{4}{12} = \dfrac{10}{12} = \dfrac{5}{6}$

9. $\dfrac{2}{6} + \dfrac{1}{4} = \dfrac{8}{24} + \dfrac{6}{24} = \dfrac{14}{24} = \dfrac{7}{12}$

10. $\dfrac{3}{4} - \dfrac{1}{5} = \dfrac{15}{20} - \dfrac{4}{20} = \dfrac{11}{20}$

11. $\dfrac{4}{5} - \dfrac{1}{2} = \dfrac{8}{10} - \dfrac{5}{10} = \dfrac{3}{10}$

12. $2\dfrac{1}{3} = \dfrac{7}{3}$

13. $1\dfrac{1}{5} = \dfrac{6}{5}$

14. $6\dfrac{1}{2} = \dfrac{13}{2}$

15. $3\dfrac{4}{5} = \dfrac{19}{5}$

16. $5\dfrac{3}{4}$ dollars $= \dfrac{23}{4}$ dollars, or 23 quarters

17. $1 \times 10^2 + 7 \times 10^1 + 6 \times 10^0 + 4 \times \dfrac{1}{10^1}$

18. grams

19.
$$
\begin{array}{r}
11.50 \\
+ 12.25 \\
\hline
23.75 \text{ minutes spent by Rachel}
\end{array}
$$

$$
\begin{array}{r}
{}^{2\ \ 9\ \ 14} \\
\not{3}\not{0}.\not{5}{}^1 0 \\
-2\ 3\ .\ 7\ 5 \\
\hline
6\ .\ 7\ 5 \text{ more minutes Sierra spent}
\end{array}
$$

20.
$$
\begin{array}{r}
8.875 \\
-6.625 \\
\hline
2.250 \text{ in}
\end{array}
$$

Systematic Review 6F

1. $\dfrac{1 \text{ kilogram(kg)}}{1,000 \text{ grams (g)}}$; $\dfrac{1 \text{ hectogram(hg)}}{100 \text{ grams (g)}}$;

$\dfrac{1 \text{ dekagram(dag)}}{10 \text{ grams (g)}}$; $\dfrac{1 \text{ gram(g)}}{1 \text{ gram(g)}}$

2. $\dfrac{1 \text{ kiloliter(kl)}}{1,000 \text{ liters (L)}}$; $\dfrac{1 \text{ hectoliter(hl)}}{100 \text{ liters (L)}}$;

$\dfrac{\text{dekaliter(dal)}}{10 \text{ liters (L)}}$; $\dfrac{1 \text{ liter(L)}}{1 \text{ liter(L)}}$

3. $\dfrac{1 \text{ kilometer(km)}}{1,000 \text{ meters (m)}}$; $\dfrac{1 \text{ hectometer(hm)}}{100 \text{ meters (m)}}$;

$\dfrac{1 \text{ dekameter(dam)}}{10 \text{ meters (m)}}$; $\dfrac{1 \text{ meter(m)}}{1 \text{ meter(m)}}$

4.
$$
\begin{array}{r}
6.11 \\
+ \ .05 \\
\hline
6.16
\end{array}
$$

5.
$$
\begin{array}{r}
4.8 \\
-1.0 \\
\hline
3.8
\end{array}
$$

6.
$$
\begin{array}{r}
{}^1 \\
3.491 \\
+4.276 \\
\hline
7.767
\end{array}
$$

7.
$$\begin{array}{r} \overset{1}{\cancel{2}}.\overset{9}{\cancel{0}}\,\overset{9}{\cancel{0}}\,7 \\ -\ .3\ 8\ 9 \\ \hline 1.6\ 1\ 8 \end{array}$$

8. $\dfrac{2}{6} + \dfrac{1}{5} = \dfrac{10}{30} + \dfrac{6}{30} = \dfrac{16}{30} = \dfrac{8}{15}$

9. $\dfrac{1}{2} + \dfrac{3}{9} = \dfrac{9}{18} + \dfrac{6}{18} = \dfrac{15}{18} = \dfrac{5}{6}$

10. $\dfrac{4}{7} - \dfrac{1}{4} = \dfrac{16}{28} - \dfrac{7}{28} = \dfrac{9}{28}$

11. $\dfrac{5}{6} - \dfrac{1}{8} = \dfrac{40}{48} - \dfrac{6}{48} = \dfrac{34}{48} = \dfrac{17}{24}$

12. $1\dfrac{1}{8} = \dfrac{9}{8}$

13. $3\dfrac{2}{5} = \dfrac{17}{5}$

14. $5\dfrac{1}{4} = \dfrac{21}{4}$

15. $7\dfrac{3}{10} = \dfrac{73}{10}$

16. $3\dfrac{5}{6} = \dfrac{23}{6}$ of a pie; 23 people

17. $5^3 = 5 \times 5 \times 5 = 125$

18.
$$\begin{array}{r} \overset{1}{\ }\,3.6 \\ +\ .8 \\ \hline 4.4 \text{ rolls needed} \end{array}$$

$$\begin{array}{r} \overset{4}{\cancel{5}}.\overset{1}{\ }0 \\ -\ 4.4 \\ \hline .6 \text{ roll left over} \end{array}$$

19. liters

20. $16 \div 4 = 4$;

$4 \times 3 = 12$ children wanted to play

$16 - 12 = 4$ children didn't want to play

Lesson Practice 7A

1. yard
2. quart
3. inch
4. f: $\dfrac{1}{10}$
5. e: $\dfrac{1}{100}$
6. d: $\dfrac{1}{1,000}$

7. $\dfrac{1 \text{ gram(g)}}{1 \text{ gram(g)}}$; $\dfrac{1 \text{ decigram(dg)}}{\frac{1}{10} \text{ gram(g)}}$;

$\dfrac{1 \text{ centigram(cg)}}{\frac{1}{100} \text{ gram(g)}}$; $\dfrac{1 \text{ milligram(mg)}}{\frac{1}{1,000} \text{ gram(g)}}$

8. $\dfrac{1 \text{ liter(L)}}{1 \text{ liter(L)}}$; $\dfrac{1 \text{ deciliter(dl)}}{\frac{1}{10} \text{ liter(L)}}$;

$\dfrac{1 \text{ centiliter(cl)}}{\frac{1}{100} \text{ liter(L)}}$; $\dfrac{1 \text{ milliliter(ml)}}{\frac{1}{1,000} \text{ liter(L)}}$

9. $\dfrac{1 \text{ meter(m)}}{1 \text{ meter(m)}}$; $\dfrac{1 \text{ decimeter(dm)}}{\frac{1}{10} \text{ meter(m)}}$;

$\dfrac{1 \text{ centimeter(cm)}}{\frac{1}{100} \text{ meter(m)}}$; $\dfrac{1 \text{ millimeter(mm)}}{\frac{1}{1,000} \text{ meter(m)}}$

10. $100 \text{ cm} = 1 \text{ m}$

11. 1 deciliter

12. $28 \text{ g} \times 16 \text{ g} = 448 \text{ grams}$

Lesson Practice 7B

1. gram
2. kilogram
3. kilometers
4. $\dfrac{1}{100}$
5. $\dfrac{1}{10}$
6. $\dfrac{1}{1,000}$

7. $\dfrac{1 \text{ gram(g)}}{1 \text{ gram(g)}}$; $\dfrac{1 \text{ decigram(dg)}}{\frac{1}{10} \text{ gram(g)}}$;

$\dfrac{1 \text{ centigram(cg)}}{\frac{1}{100} \text{ gram(g)}}$; $\dfrac{1 \text{ milligram(mg)}}{\frac{1}{1,000} \text{ gram(g)}}$

8. $\dfrac{1 \text{ liter(L)}}{1 \text{ liter(L)}}$; $\dfrac{1 \text{ deciliter(dl)}}{\frac{1}{10} \text{ liter(L)}}$;

$\dfrac{1 \text{ centiliter(cl)}}{\frac{1}{100} \text{ liter(L)}}$; $\dfrac{1 \text{ milliliter(ml)}}{\frac{1}{1,000} \text{ liter(L)}}$

9. $\dfrac{1\ \text{meter(m)}}{1\ \text{meter(m)}}$; $\dfrac{1\ \text{decimeter(dm)}}{\frac{1}{10}\ \text{meter(m)}}$;

$\dfrac{1\ \text{centimeter(cm)}}{\frac{1}{100}\ \text{meter(m)}}$; $\dfrac{1\ \text{millimeter(mm)}}{\frac{1}{1,000}\ \text{meter(m)}}$

10. 1 milliliter
11. 1 centimeter
12. grams

Lesson Practice 7C

1. meters
2. 2 miles
3. milliliters
4. grams
5. 16 ounces
6. 36 inches
7. $\dfrac{1\ \text{gram(g)}}{1\ \text{gram(g)}}$; $\dfrac{1\ \text{decigram(dg)}}{\frac{1}{10}\ \text{gram(g)}}$;

$\dfrac{1\ \text{centigram(cg)}}{\frac{1}{100}\ \text{gram(g)}}$; $\dfrac{1\ \text{milligram(mg)}}{\frac{1}{1,000}\ \text{gram(g)}}$

8. $\dfrac{1\ \text{liter(L)}}{1\ \text{liter(L)}}$; $\dfrac{1\ \text{deciliter(dl)}}{\frac{1}{10}\ \text{liter(L)}}$;

$\dfrac{1\ \text{centiliter(cl)}}{\frac{1}{100}\ \text{liter(L)}}$; $\dfrac{1\ \text{milliliter(ml)}}{\frac{1}{1,000}\ \text{liter(L)}}$

9. $\dfrac{1\ \text{meter(m)}}{1\ \text{meter(m)}}$; $\dfrac{1\ \text{decimeter(dm)}}{\frac{1}{10}\ \text{meter(m)}}$;

$\dfrac{1\ \text{centimeter(cm)}}{\frac{1}{100}\ \text{meter(m)}}$; $\dfrac{1\ \text{millimeter(mm)}}{\frac{1}{1,000}\ \text{meter(m)}}$

10. 1 millimeter
11. 1 liter
12. 50 grams

Systematic Review 7D

1. $\dfrac{1\ \text{gram(g)}}{1\ \text{gram(g)}}$; $\dfrac{1\ \text{decigram(dg)}}{\frac{1}{10}\ \text{gram(g)}}$;

$\dfrac{1\ \text{centigram(cg)}}{\frac{1}{100}\ \text{gram(g)}}$; $\dfrac{1\ \text{milligram(mg)}}{\frac{1}{1,000}\ \text{gram(g)}}$

2. $\dfrac{1\ \text{liter(L)}}{1\ \text{liter(L)}}$; $\dfrac{1\ \text{deciliter(dl)}}{\frac{1}{10}\ \text{liter(L)}}$;

$\dfrac{1\ \text{centiliter(cl)}}{\frac{1}{100}\ \text{liter(L)}}$; $\dfrac{1\ \text{milliliter(ml)}}{\frac{1}{1,000}\ \text{liter(L)}}$

3. $\dfrac{1\ \text{meter(m)}}{1\ \text{meter(m)}}$; $\dfrac{1\ \text{decimeter(dm)}}{\frac{1}{10}\ \text{meter(m)}}$;

$\dfrac{1\ \text{centimeter(cm)}}{\frac{1}{100}\ \text{meter(m)}}$; $\dfrac{1\ \text{millimeter(mm)}}{\frac{1}{1,000}\ \text{meter(m)}}$

4. c: 1,000
5. b: 100
6. a: 10
7. $16 \div 2 = 8$
   $8 \times 1 = 8$
8. $40 \div 5 = 8$
   $8 \times 3 = 24$
9. $63 \div 9 = 7$
   $7 \times 2 = 14$
10. $48 \div 4 = 12$
    $12 \times 1 = 12$
11. $\dfrac{25}{8} = 3\dfrac{1}{8}$
12. $\dfrac{3}{2} = 1\dfrac{1}{2}$
13. $\dfrac{11}{9} = 1\dfrac{2}{9}$
14. $\dfrac{10}{3} = 3\dfrac{1}{3}$
15. liter
16. meter
17. $\begin{array}{r} {\scriptstyle 1} \\ 7.2 \\ +\ 6.3 \\ \hline 13.5 \end{array}$ tons brought by the first two trucks

$\begin{array}{r} 17.8\,0 \\ -13.5\,0 \\ \hline 4.3\,0 \end{array}$ tons needed

18. kilometers

Systematic Review 7E

1. $\dfrac{1 \text{ gram(g)}}{1 \text{ gram(g)}}$; $\dfrac{1 \text{ decigram(dg)}}{\frac{1}{10} \text{ gram(g)}}$;

   $\dfrac{1 \text{ centigram(cg)}}{\frac{1}{100} \text{ gram(g)}}$; $\dfrac{1 \text{ milligram(mg)}}{\frac{1}{1,000} \text{ gram(g)}}$

2. $\dfrac{1 \text{ liter(L)}}{1 \text{ liter(L)}}$; $\dfrac{1 \text{ deciliter(dl)}}{\frac{1}{10} \text{ liter(L)}}$;

   $\dfrac{1 \text{ centiliter(cl)}}{\frac{1}{100} \text{ liter(L)}}$; $\dfrac{1 \text{ milliliter(ml)}}{\frac{1}{1,000} \text{ liter(L)}}$

3. $\dfrac{1 \text{ meter(m)}}{1 \text{ meter(m)}}$; $\dfrac{1 \text{ decimeter(dm)}}{\frac{1}{10} \text{ meter(m)}}$;

   $\dfrac{1 \text{ centimeter(cm)}}{\frac{1}{100} \text{ meter(m)}}$; $\dfrac{1 \text{ millimeter(mm)}}{\frac{1}{1,000} \text{ meter(m)}}$

4. deka
5. hecto
6. kilo
7. $30 \div 3 = 10$
   $10 \times 1 = 10$
8. $12 \div 6 = 2$
   $2 \times 5 = 10$
9. $49 \div 7 = 7$
   $7 \times 1 = 7$
10. $64 \div 8 = 8$
    $8 \times 5 = 40$
11. $\dfrac{37}{9} = 4\dfrac{1}{9}$
12. $\dfrac{17}{4} = 4\dfrac{1}{4}$
13. $\dfrac{47}{10} = 4\dfrac{7}{10}$
14. $\dfrac{53}{5} = 10\dfrac{3}{5}$
15. centimeters
16. 1,000
17. 1 mile
18. kilogram
19. $\begin{array}{r} 2.50 \\ +3.25 \\ \hline 5.75 \text{ gallons used} \end{array}$

    $\begin{array}{r} {}^{6}\cancel{7}.{}^{14}\cancel{5}{}^{1}0 \\ -5.75 \\ \hline 1.75 \text{ gallons left} \end{array}$
20. $\dfrac{1}{2} - \dfrac{1}{6} = \dfrac{6}{12} - \dfrac{2}{12} = \dfrac{4}{12} = \dfrac{1}{3}$ of the job

Systematic Review 7F

1. $\dfrac{1 \text{ gram(g)}}{1 \text{ gram(g)}}$; $\dfrac{1 \text{ decigram(dg)}}{\frac{1}{10} \text{ gram(g)}}$;

   $\dfrac{1 \text{ centigram(cg)}}{\frac{1}{100} \text{ gram(g)}}$; $\dfrac{1 \text{ milligram(mg)}}{\frac{1}{1,000} \text{ gram(g)}}$

2. $\dfrac{1 \text{ liter(L)}}{1 \text{ liter(L)}}$; $\dfrac{1 \text{ deciliter(dl)}}{\frac{1}{10} \text{ liter(L)}}$;

   $\dfrac{1 \text{ centiliter(cl)}}{\frac{1}{100} \text{ liter(L)}}$; $\dfrac{1 \text{ milliliter(ml)}}{\frac{1}{1,000} \text{ liter(L)}}$

3. $\dfrac{1 \text{ meter(m)}}{1 \text{ meter(m)}}$; $\dfrac{1 \text{ decimeter(dm)}}{\frac{1}{10} \text{ meter(m)}}$;

   $\dfrac{1 \text{ centimeter(cm)}}{\frac{1}{100} \text{ meter(m)}}$; $\dfrac{1 \text{ millimeter(mm)}}{\frac{1}{1,000} \text{ meter(m)}}$

4. dekagram
5. hectoliter
6. kilometer
7. $72 \div 8 = 9$
   $9 \times 3 = 27$
8. $50 \div 2 = 25$
   $25 \times 1 = 25$
9. $24 \div 3 = 8$
   $8 \times 2 = 16$
10. $44 \div 11 = 4$
    $4 \times 4 = 16$
11. $\dfrac{28}{6} = 4\dfrac{4}{6} = 4\dfrac{2}{3}$
12. $\dfrac{11}{7} = 1\dfrac{4}{7}$
13. $\dfrac{16}{5} = 3\dfrac{1}{5}$

14. $\dfrac{31}{2} = 15\dfrac{1}{2}$

15. the meter stick

16. 10 pounds

17. liters

18. $\begin{array}{r} \overset{1}{}\overset{1}{} \\ \$\;6.15 \\ \$\;8.45 \\ +\$\,10.25 \\ \hline \$\,24.85 \end{array}$

19. meters

20. $780 \div 10 = 78$
    $78 \times 3 = 234$ people

Lesson Practice 8A

1. done

2. $1\ \cancel{\text{dam}} \times \dfrac{10\ \cancel{\text{m}}}{1\ \cancel{\text{dam}}} \times \dfrac{100\text{cm}}{1\ \cancel{\text{m}}} = 1{,}000$ cm

   Or three steps, so $1(10)(10)(10) = 1{,}000$ cm

3. 10 ml

4. 1,000 dl

5. 1,000 mg

6. 100 mg

7. done

8. 3,000,000 mm

9. 3,500 ml

10. 1,300 g

11. 600,000 cl

12. 1,000 dg

13. 60,000 cg

14. 100,000 dm

15. 90 pieces

Lesson Practice 8B

1. 10,000 mm

2. 1,000,000 ml

3. 10 cl

4. 10 hg

5. 1,000 dm

6. 100 mm

7. 80,000 cm

8. 2,000 cl

9. 6,000,000 mg

10. 3,200 dg

11. 90 hl

12. 2,200,000 cm

13. 3,000 ml

14. 4,500 dg

15. 1,200 cm

Lesson Practice 8C

1. 100,000 cm

2. 10 dam

3. 10,000 mg

4. 10,000 cg

5. 10,000 dl

6. 10 L

7. 1,000 dam

8. 150,000 cg

9. 700 cm

10. 60 mm

11. 5,000 ml

12. 24,000 g

13. 500,000 seconds

14. 80 pieces

15. 2 liters = 2,000 ml
    2,000 ml > 1,983 ml
    2 liters is more

Systematic Review 8D

1. 500 cg

2. 10,000 mm

3. 160,000 cm

4. c: 1,000

5. b: 100

6. a: 10

7. d: $\dfrac{1}{10}$

8. f: $\dfrac{1}{100}$

9. e: $\dfrac{1}{1{,}000}$

10. $2\dfrac{3}{8} = \dfrac{19}{8}$

11. $6\dfrac{1}{7} = \dfrac{43}{7}$

12. $\dfrac{17}{9} = 1\dfrac{8}{9}$

13. $\dfrac{8}{5} = 1\dfrac{3}{5}$

14. done

15. $3\dfrac{3}{8} + 1\dfrac{4}{5} = 3\dfrac{15}{40} + 1\dfrac{32}{40} = 4\dfrac{47}{40} =$

$4 + \dfrac{40}{40} + \dfrac{7}{40} = 4 + 1 + \dfrac{7}{40} = 5\dfrac{7}{40}$

16. $2\dfrac{1}{10} + 3\dfrac{5}{8} = 2\dfrac{8}{80} + 3\dfrac{50}{80} = 5\dfrac{58}{80} = 5\dfrac{29}{40}$

17. 2,000 g

18. $1\dfrac{1}{2}$ hr $+ 1\dfrac{1}{4}$ hr $= 1\dfrac{4}{8}$ hr $+ 1\dfrac{2}{8}$ hr

$= 2\dfrac{6}{8}$ hr $= 2\dfrac{3}{4}$ hr

19.
$$\begin{array}{r} {}^{3}\cancel{4}\,{}^{10}\cancel{1}.{}^{14}\cancel{5}\,{}^{1}0 \\ -\ 3\ 7.\ 7\ 5 \\ \hline 3.\ 7\ 5\ \text{in} \end{array}$$

20. $\dfrac{1}{3}$ of 900 = 300 in tanks

$\dfrac{1}{6}$ of 900 = 150 in trucks

300 + 150 = 450 riding

900 − 450 = 450 walking

You could also have added the fractions and used your answer to find the number of soldiers who were riding. There is often more than one way to solve a word problem.

Systematic Review 8E

1. 32,000 ml
2. 90,000 ml
3. 150 mm
4. b: 1 yard
5. a: 0.4 of an inch
6. c: 0.6 of a mile
7. e: 1 quart
8. f: $\dfrac{1}{500}$ of a pound
9. d: 2.2 pounds
10. $5\dfrac{1}{8} = \dfrac{41}{8}$
11. $4\dfrac{2}{3} = \dfrac{14}{3}$
12. $\dfrac{22}{7} = 3\dfrac{1}{7}$
13. $\dfrac{27}{5} = 5\dfrac{2}{5}$
14. $8\dfrac{2}{3} + 5\dfrac{1}{4} = 8\dfrac{8}{12} + 5\dfrac{3}{12} = 13\dfrac{11}{12}$
15. $18\dfrac{1}{10} + 3\dfrac{5}{8} = 18\dfrac{8}{80} + 3\dfrac{50}{80} =$

$21\dfrac{58}{80} = 21\dfrac{29}{40}$
16. $11\dfrac{3}{5} + 4\dfrac{1}{6} = 11\dfrac{18}{30} + 4\dfrac{5}{30} = 15\dfrac{23}{30}$
17. 100 cm = 10 dm; they are the same length
18. 1 km

Systematic Review 8F

1. 400,000 cg
2. 800 dag
3. 5,000,000 ml
4. b: length
5. c: volume
6. a: weight
7. d: length
8. f: weight
9. e: volume
10. $2\dfrac{3}{5} = \dfrac{13}{5}$
11. $1\dfrac{1}{7} = \dfrac{8}{7}$
12. $\dfrac{19}{8} = 2\dfrac{3}{8}$
13. $\dfrac{31}{6} = 5\dfrac{1}{6}$
14. $9\dfrac{1}{3} + 6\dfrac{1}{4} = 9\dfrac{4}{12} + 6\dfrac{3}{12} = 15\dfrac{7}{12}$
15. $4\dfrac{2}{3} + 1\dfrac{1}{5} = 4\dfrac{10}{15} + 1\dfrac{3}{15} = 5\dfrac{13}{15}$
16. $12\dfrac{2}{10} + 4\dfrac{5}{8} = 12\dfrac{16}{80} + 4\dfrac{50}{80} =$

$16\dfrac{66}{80} = 16\dfrac{33}{40}$
17. 45,000 g

45,000,000 mg
18. $5 \times 10 = 50$ bottles

19. $1 \times 10^1 + 2 \times 10^0 + 8 \times \dfrac{1}{10^1} +$

$7 \times \dfrac{1}{10^2} + 4 \times \dfrac{1}{10^3}$

20.
```
    1
  6.20
  4.45
+ 9.00
─────────
 19.65 ft
```

Lesson Practice 9A

1. done

2. done

3.
```
   1+.4        1.4
 × 1+.2      ×1.2
 ─────────   ─────
   .2+.08      .28
   1+.4        1.4
 ─────────   ─────
 1+.6+.08    (1.68)
```

4.
```
   1+ .1       1.1
 ×   .6      ×  .6
 ─────────   ─────
   .6+.06     (.66)
```

5.
```
   1+.2        1.2
 ×2+.1       ×2.1
 ─────────   ─────
   .1+.02      .12
   2+.4        2.4
 ─────────   ─────
 2+.5+.02    (2.52)
```

6.
```
   2+.2        2.2
 ×   .3      ×  .3
 ─────────   ─────
   .6+.06     (.66)
```

7.
```
   1+.3        1.3
 × 1+.1      ×1.1
 ─────────   ─────
   .1+.03      .13
   1+.3        1.3
 ─────────   ─────
 1+.4+.03    (1.43)
```

8.
```
   2+.3        2.3
 ×   .2      ×  .2
 ─────────   ─────
   .4+.06     (.46)
```

9.
```
   3.1
 ×  .3
 ─────────
  (.93 yd)
```

10.
```
   2.1
 ×  .4
 ─────────
  (.84 gal)
```

Lesson Practice 9B

1.
```
   2+.8        2.8
 ×1+.0       ×1.0
 ─────────   ─────
   .0+.00      .00
   2+.8        2.8
 ─────────   ─────
 2+.8+.00    (2.80)
```

2.
```
   2+.3        2.3
 ×   .1      ×  .1
 ─────────   ─────
   .2+.03     (.23)
```

3.
```
   2+.1        2.1
 ×1+.4       ×1.4
 ─────────   ─────
   .8+.04      .84
   2+.1        2.1
 ─────────   ─────
 2+.9+.04    (2.94)
```

4.
```
   1+.1        1.1
 ×   .8      ×  .8
 ─────────   ─────
   .8+.08     (.88)
```

5.
```
   2+.2        2.2
 ×1+.2       ×1.2
 ─────────   ─────
   .4+.04      .44
   2+.2        2.2
 ─────────   ─────
 2+.6+.04    (2.64)
```

6.
```
   1+.3        1.3
 ×   .3      ×  .3
 ─────────   ─────
   .3+.09     (.39)
```

7.
```
   1+.3        1.3
 ×1+.3       ×1.3
 ─────────   ─────
   .3+.09      .39
   1+.3        1.3
 ─────────   ─────
 1+.6+.09    (1.69)
```

8.
```
   1+.1        1.1
 ×   .9      ×  .9
 ─────────   ─────
   .9+.09     (.99)
```

9.
```
   2.6
 ×1.1
 ─────────
   .26
   2.6
 ─────────
  (2.86)
```

10.
```
   1.1
 ×  .5
 ─────────
  (.55 mi)
```

Lesson Practice 9C

1.
```
  1 + .3           1.3
× 2 + .2         × 2.2
   .2 + .06        .26
  2 + .6          2.6
  2 + .8 + .06    2.86
```

2.
```
  2 + .0          2.0
×    .1         ×  .1
   .2 + .00       .2
```

3.
```
  1 + .3           1.3
× 1 + .2         × 1.2
   .2 + .06        .26
  1 + .3          1.3
  1 + .5 + .06    1.56
```

4.
```
  2 + .1          2.1
×    .6         ×  .6
  1 + .2 + .06   1.26
```

5.
```
  2 + .1           2.1
× 1 + .4         × 1.4
   .8 + .04        .84
  2 + .1          2.1
  2 + .9 + .04    2.94
```

6.
```
  2 + .1          2.1
×    .4         ×  .4
   .8 + .04        .84
```

7.
```
  2 + .3           2.3
× 1 + .1         × 1.1
   .2 + .03        .23
  2 + .3          2.3
  2 + .5 + .03    2.53
```

8.
```
   .3            .3
×  .3         ×  .3
   .09           .09
```

9.
```
3.2 miles/hour
×  .1 hour
   .32 miles
```

10.
```
   .4
×  .2
   .08 of the jelly beans
```

Systematic Review 9D

1.
```
  2.5
× 1.1
   .25
  2.5
  2.75
```

2.
```
  2.3
× 1.3
   .69
  2.3
  2.99
```

3.
```
  1.1
×  .5
   .55
```

4.
```
   .2
×  .2
   .04
```

5. 900 mg
6. 8,000 cl
7. 2,400 cg
8. cg
9. ml
10. km
11. dal
12. dl
13. kg
14. done
15. done

16. $5\frac{1}{4} - 1\frac{2}{3} = 5\frac{3}{12} - 1\frac{8}{12} = 4\frac{15}{12} - 1\frac{8}{12} = 3\frac{7}{12}$

17. $3\frac{1}{2}$ yd $- \frac{1}{3}$ yd $= 3\frac{3}{6}$ yd $- \frac{2}{6}$ yd $= 3\frac{1}{6}$ yards

18.
```
  3.3
×  .3
   .99 pies eaten
```

```
  2  12
  3. 3  0
-    .9 9
  2.3 1 pies left over
```

Systematic Review 9E

1. $\begin{array}{r} 2.4 \\ \times\ .2 \\ \hline .48 \end{array}$

2. $\begin{array}{r} 1.8 \\ \times 1.1 \\ \hline .18 \\ 1.8 \\ \hline 1.98 \end{array}$

3. $\begin{array}{r} 1.7 \\ \times\ .1 \\ \hline .17 \end{array}$

4. $\begin{array}{r} .1 \\ \times .1 \\ \hline .01 \end{array}$

5. 280,000 cm

6. 3,600 dm

7. 500 dal

8. meter

9. quart

10. gram

11. $7\frac{2}{5} - 3\frac{1}{3} = 7\frac{6}{15} - 3\frac{5}{15} = 4\frac{1}{15}$

12. $9 - 2\frac{1}{2} = 8\frac{2}{2} - 2\frac{1}{2} = 6\frac{1}{2}$

13. $6\frac{1}{2} - 4\frac{1}{7} = 6\frac{7}{14} - 4\frac{2}{14} = 2\frac{5}{14}$

14. $\$7\frac{1}{2} - \$1\frac{1}{4} = \$7\frac{2}{4} - \$1\frac{1}{4}$

    $= 6\frac{1}{4}$ dollars or $6.25

    If one fraction can be made into an equivalent fraction with the same denominator as the other, you don't have to use the rule of four. The final result is the same.

15. $\begin{array}{r} 4.4 \text{ hours} \\ \times\ .1 \text{ pounds/hour} \\ \hline .44 \text{ pounds} \end{array}$

16. $\begin{array}{r} {\scriptstyle 1\ \ 1} \\ 45.15 \\ 34.10 \\ +\ 7.05 \\ \hline 86.30 \text{ pounds} \end{array}$

17. $6\ \cancel{hm} \times \dfrac{100\ \cancel{m}}{1\ \cancel{hm}} \times \dfrac{10\ dm}{1\ \cancel{m}} = 6,000\ dm$

    so she took 6,000 steps.

18. A meter is a little longer than a yard, so he ran faster in the 100 meter dash.

Systematic Review 9F

1. $\begin{array}{r} 2.3 \\ \times\ .3 \\ \hline .69 \end{array}$

2. $\begin{array}{r} 2.1 \\ \times 1.3 \\ \hline .63 \\ 2.1 \\ \hline 2.73 \end{array}$

3. $\begin{array}{r} 2.2 \\ \times\ .4 \\ \hline .88 \end{array}$

4. $\begin{array}{r} .4 \\ \times .2 \\ \hline .08 \end{array}$

5. 140 mm

6. 2,000 g

7. 110 m

8. pounds

9. ounce

10. kilometer

11. $10\frac{1}{3} - 4\frac{1}{4} = 10\frac{4}{12} - 4\frac{3}{12} = 6\frac{1}{12}$

12. $6 - 1\frac{2}{3} = 5\frac{3}{3} - 1\frac{2}{3} = 4\frac{1}{3}$

13. $13\frac{1}{5} - 8\frac{5}{6} = 13\frac{6}{30} - 8\frac{25}{30} =$

    $12\frac{36}{30} - 8\frac{25}{30} = 4\frac{11}{30}$

14. $1\ \cancel{L} \times \dfrac{1,000\ ml}{1\ \cancel{L}} = 1,000\ ml$

    so about 1,000 g

15. $1,000\ \cancel{g} \times \dfrac{1\ kg}{1,000\ \cancel{g}} = 1\ kg$

16. $\begin{array}{r} 1.1 \text{ pounds} \\ \times\ .5 \\ \hline .55 \text{ pounds} \end{array}$

17. $8 \text{ kg} \times \dfrac{1{,}000 \text{ g}}{1 \text{ kg}} = 8{,}000 \text{ g}$

18.
```
      6
    ⁷.¹0
  − 2 . 3
  ─────────
    4 . 7  minutes
```

Lesson Practice 10A

1. done
2. done
3.
```
      35.                35   0 places
   ×  .26             ×  26   2 places
   ───────            ───────
     2.10               2 10
     7.0                70
   ───────            ───────
   ( 9.10 )           9.10    2 places
```
4.
```
     1.03              103   2 places
   ×  .76            ×  76   2 places
   ───────           ───────
    .0618              6 18
    .721              72 1
   ───────           ───────
   (.7828)           .7828   4 places
```
5.
```
     4.7               47    1 place
   ×  .6             ×  6    1 place
   ───────           ───────
   ( 2.82 )          2.82    2 places
```
6.
```
     .52               52    2 places
   ×.14              ×  14   2 places
   ───────           ───────
    .0208              208
    .052              52
   ───────           ───────
   (.0728)           .0728   4 places
```
7.
```
     200.              200   0 places
   ×   .08           ×  08   2 places
   ───────           ───────
   ( 16.00 )         16.00   2 places
```
8.
```
     5.28              528   2 places
   ×  .12            ×  12   2 places
   ───────           ───────
    .1056             1056
    .528              528
   ───────           ───────
   (.6336)           .6336   4 places
```
9. $\$6.55/\text{hr} \times 0.4 \text{ hr} = \$2.62$
10. $\$0.96/\text{doz} \times 2.5 \text{ doz} = \$2.40$
    (Note: The final zero is
    included for money.)

Lesson Practice 10B

1.
```
      3.4               34   1 place
   ×  .94            ×  94   2 places
   ───────           ───────
    .136               136
    3.06              306
   ───────           ───────
   ( 3.196 )         3.196   3 places
```
2.
```
     .73                73   2 places
   ×.48              ×  48   2 places
   ───────           ───────
    .0584             0584
    .292              292
   ───────           ───────
   (.3504)           .3504   4 places
```
3.
```
      33.               33   0 places
   ×  .42            ×  42   2 places
   ───────           ───────
    .66                66
    13.2              13 2
   ───────           ───────
   ( 13.86 )         13.86   2 places
```
4.
```
     5.19              5 19  2 places
   ×  .81            ×   81  2 places
   ───────           ───────
    .0519              5 19
    4.152             4 152
   ───────           ───────
   ( 4.2039 )        4.2039  4 places
```
5.
```
     9.6                96   1 place
   ×  .9             ×   9   1 place
   ───────           ───────
   ( 8.64 )          8.64    2 places
```
6.
```
     .64                64   2 places
   ×.94              ×   94  2 places
   ───────           ───────
    .0256              256
    .576              576
   ───────           ───────
   (.6016)           .60 16  4 places
```
7.
```
      116.             116   0 places
   ×   .02           ×  02   2 places
   ───────           ───────
   ( 2.32 )          2.32    2 places
```
8.
```
     7.18              718   2 places
   ×  .05            ×   5   2 places
   ───────           ───────
   (.3590)           .3590   4 places
```
9. $0.15 \text{ oz} \times 100 = 15 \text{ ounces}$
10. $0.33 \text{ g} \times 45 = 14.85 \text{ grams}$

Lesson Practice 10C

1.

$$\begin{array}{r} 4.8 \\ \times\ .71 \\ \hline .048 \\ 3.36 \\ \hline 3.408 \end{array}$$

$$\begin{array}{r} 48\ \text{1 place} \\ \times\ 71\ \text{2 places} \\ \hline 048 \\ 3\ 36 \\ \hline 3.408\ \text{3 places} \end{array}$$

2.

$$\begin{array}{r} .62 \\ \times .37 \\ \hline .0434 \\ .186 \\ \hline .2294 \end{array}$$

$$\begin{array}{r} 62\ \text{2 places} \\ \times\ 37\ \text{2 places} \\ \hline 434 \\ 186 \\ \hline .2294\ \text{4 places} \end{array}$$

3.

$$\begin{array}{r} 69. \\ \times\ 2.3 \\ \hline 20.7 \\ 138 \\ \hline 158.7 \end{array}$$

$$\begin{array}{r} 6\ 9\ \text{0 places} \\ \times\ 2\ 3\ \text{1 place} \\ \hline 20\ 7 \\ 138 \\ \hline 158.7\ \text{1 place} \end{array}$$

4.

$$\begin{array}{r} 9.97 \\ \times\ .11 \\ \hline .0997 \\ .997 \\ \hline 1.0967 \end{array}$$

$$\begin{array}{r} 997\ \text{2 places} \\ \times\ 11\ \text{2 places} \\ \hline 997 \\ 997 \\ \hline 1.0967\ \text{4 places} \end{array}$$

5.

$$\begin{array}{r} 1.7 \\ \times\ .4 \\ \hline .68 \end{array}$$

$$\begin{array}{r} 17\ \text{1 place} \\ \times\ 4\ \text{1 place} \\ \hline 68\ \text{2 places} \end{array}$$

6.

$$\begin{array}{r} .16 \\ \times .54 \\ \hline .0064 \\ .080 \\ \hline .0864 \end{array}$$

$$\begin{array}{r} 16\ \text{2 places} \\ \times\ 54\ \text{2 places} \\ \hline 64 \\ 80 \\ \hline .0864\ \text{4 places} \end{array}$$

7.

$$\begin{array}{r} 400. \\ \times\ .11 \\ \hline 44.00 \end{array}$$

$$\begin{array}{r} 4\ 00\ \text{0 places} \\ \times\ 11\ \text{2 places} \\ \hline 44.00\ \text{2 places} \end{array}$$

8.

$$\begin{array}{r} 6.73 \\ \times\ .46 \\ \hline .4038 \\ 2.692 \\ \hline 3.0958 \end{array}$$

$$\begin{array}{r} 673\ \text{2 places} \\ \times\ 46\ \text{2 places} \\ \hline 4038 \\ 2\ 692 \\ \hline 3.0958\ \text{4 places} \end{array}$$

9. $4.25\ \text{bushels} \times 0.75 = 3.1875\ \text{bushel}$

10. $2.45\ \text{meters} \times 5 = 12.25\ \text{meters}$

Systematic Review 10D

1.

$$\begin{array}{r} 2.6 \\ \times\ .24 \\ \hline .104 \\ .52 \\ \hline .624 \end{array}$$

$$\begin{array}{r} 26\ \text{1 place} \\ \times\ 24\ \text{2 places} \\ \hline 104 \\ 52 \\ \hline .624\ \text{3 places} \end{array}$$

2.

$$\begin{array}{r} 6.3 \\ \times\ 5.7 \\ \hline 4.41 \\ 31.5 \\ \hline 35.91 \end{array}$$

$$\begin{array}{r} 63\ \text{1 place} \\ \times\ 57\ \text{1 place} \\ \hline 4\ 41 \\ 31\ 5 \\ \hline 35.91\ \text{2 places} \end{array}$$

3.

$$\begin{array}{r} 3.52 \\ \times\ .04 \\ \hline .1408 \end{array}$$

$$\begin{array}{r} 352\ \text{2 places} \\ \times\ 04\ \text{2 places} \\ \hline .1408\ \text{4 places} \end{array}$$

4.

$$\begin{array}{r} .67 \\ \times .05 \\ \hline .0335 \end{array}$$

$$\begin{array}{r} 67\ \text{2 places} \\ \times\ 05\ \text{2 places} \\ \hline .0335\ \text{4 places} \end{array}$$

5. 14,000 dg

6. 50 cm

7. $1.00 - 0.12 = 0.88$

8. $4.08 - 2.9 = 1.18$

9. $0.95 + 3.61 = 4.56$

10. $4\frac{3}{4} + 2\frac{1}{5} = 4\frac{15}{20} + 2\frac{4}{20} = 6\frac{19}{20}$

11. $12\frac{4}{7} - 6\frac{1}{7} = 6\frac{3}{7}$

12. $5\frac{1}{10} + 3\frac{2}{5} = 5\frac{5}{50} + 3\frac{20}{50} =$

$8\frac{25}{50} = 8\frac{1}{2}$

or

$5\frac{1}{10} + 3\frac{4}{10} = 8\frac{5}{10} = 8\frac{1}{2}$

13. $\frac{1}{\cancel{2}} \times \frac{\cancel{2}}{3} = \frac{1}{3}$

14. $\frac{3}{4} \times \frac{1}{\cancel{9}_3} = \frac{1}{12}$

15. $\frac{\cancel{6}^3}{8} \times \frac{1}{\cancel{2}} = \frac{3}{8}$

16. $\frac{\cancel{2}}{3} \times \frac{1}{\cancel{2}} = \frac{1}{3}$ of the package

Writing "1" when canceling
is optional.

17. $0.4\ \text{in/cm} \times 18\ \text{cm} = 7.2\ \text{in}$

18. 55 kg × 2.2 lb/kg = 121 lb

Systematic Review 10E

1.

$$\begin{array}{r} 11.9 \\ \times\ 5.3 \\ \hline 3.57 \\ 59.5 \\ \hline 63.07 \end{array}$$

$$\begin{array}{r} 1\ 19 \quad \text{1 place} \\ \times\ \ 53 \quad \text{1 place} \\ \hline 3\ 57 \\ 59\ 5 \\ \hline 63.07 \quad \text{2 places} \end{array}$$

2.

$$\begin{array}{r} .05 \\ \times\ 4 \\ \hline .02 \end{array}$$

$$\begin{array}{r} 05 \quad \text{2 places} \\ \times\ \ 4 \quad \text{1 place} \\ \hline .020 \quad \text{3 places} \end{array}$$

3.

$$\begin{array}{r} 1.07 \\ \times\ .72 \\ \hline .0214 \\ 749 \\ \hline .7704 \end{array}$$

$$\begin{array}{r} 107 \quad \text{2 places} \\ \times\ \ 72 \quad \text{2 places} \\ \hline 214 \\ 749 \\ \hline .7704 \quad \text{4 places} \end{array}$$

4.

$$\begin{array}{r} .46 \\ \times .49 \\ \hline .0414 \\ .184 \\ \hline .2254 \end{array}$$

$$\begin{array}{r} 46 \quad \text{2 places} \\ \times\ \ 49 \quad \text{2 places} \\ \hline 414 \\ 184 \\ \hline .2254 \quad \text{4 places} \end{array}$$

5. 1,700 dg

6. 800,000 mm

7. 12.51 − 2.74 = 9.77

8. 2.38 + 4.509 = 6.889

9. 0.412 − 0.367 = 0.045

10. $5\frac{3}{7} + 5\frac{1}{4} = 5\frac{12}{28} + 5\frac{7}{28} = 10\frac{19}{28}$

11. $2\frac{1}{8} - 1\frac{5}{9} = 2\frac{9}{72} - 1\frac{40}{72} = 1\frac{81}{72} - 1\frac{40}{72} = \frac{41}{72}$

12. $3\frac{1}{2} + 2\frac{1}{2} = 5\frac{2}{2} = 6$

13. $\frac{1}{2} \times \frac{3}{5} = \frac{3}{10}$

14. $\frac{3}{4} \times \frac{1}{7} = \frac{3}{28}$

15. $\frac{\overset{2}{\cancel{4}}}{5} \times \frac{7}{\underset{5}{\cancel{10}}} = \frac{14}{25}$

16. $\frac{3}{4} \div \frac{6}{1} = \frac{3}{4} \times \frac{1}{\underset{2}{\cancel{6}}} = \frac{1}{8}$

17. $1\frac{1}{6}$ yd $+ 5\frac{3}{8}$ yd $= 1\frac{8}{48}$ yd $+ 5\frac{18}{48}$ yd

$= 6\frac{26}{48}$ yd $= 6\frac{13}{24}$ yd

18. 2.5 cm/in × 36 in = 90 cm

19. 1.06 qt/L × 8 L = 8.48 qt

20. 34,000 grams

Systematic Review 10F

1.

$$\begin{array}{r} 9.35 \\ \times\ .5 \\ \hline 4.675 \end{array}$$

$$\begin{array}{r} 9\ 35 \quad \text{2 places} \\ \times\ \ 5 \quad \text{1 place} \\ \hline 4.675 \quad \text{3 places} \end{array}$$

2.

$$\begin{array}{r} 8.9 \\ \times\ 8.9 \\ \hline 8.01 \\ 71.2 \\ \hline 79.21 \end{array}$$

$$\begin{array}{r} 89 \quad \text{1 place} \\ \times\ \ 89 \quad \text{1 place} \\ \hline 8\ 01 \\ 712 \\ \hline 79.21 \quad \text{2 places} \end{array}$$

3.

$$\begin{array}{r} 6.31 \\ \times\ .73 \\ \hline .1893 \\ 4.417 \\ \hline 4.6063 \end{array}$$

$$\begin{array}{r} 631 \quad \text{2 places} \\ \times\ \ 73 \quad \text{2 places} \\ \hline 1893 \\ 4\ 417 \\ \hline 4.6063 \quad \text{4 places} \end{array}$$

4.

$$\begin{array}{r} .29 \\ \times .34 \\ \hline .0116 \\ .087 \\ \hline .0986 \end{array}$$

$$\begin{array}{r} 29 \quad \text{2 places} \\ \times\ \ 34 \quad \text{2 places} \\ \hline 116 \\ 87 \\ \hline .0986 \quad \text{4 places} \end{array}$$

5. 40 dal

6. 30 L

7. 1.26 − 0.63 = 0.63

8. 3.052 + 1.98 = 5.032

9. 400 − 0.5 = 399.5

10. $6\frac{3}{6} + 8\frac{2}{5} = 6\frac{15}{30} + 8\frac{12}{30} =$

$14\frac{27}{30} = 14\frac{9}{10}$

11. $3 - 2\frac{1}{4} = 2\frac{4}{4} - 2\frac{1}{4} = \frac{3}{4}$

12. $7\frac{3}{10} + 9\frac{4}{5} = 7\frac{15}{50} + 9\frac{40}{50} =$

$16\frac{55}{50} = 17\frac{5}{50} = 17\frac{1}{10}$

13. $\frac{3}{\underset{2}{\cancel{8}}} \times \frac{\cancel{4}}{7} = \frac{3}{14}$

14. $\frac{1}{2} \times \frac{\overset{4}{\cancel{8}}}{9} = \frac{4}{9}$

15. $\frac{3}{5} \times \frac{3}{5} = \frac{9}{25}$

16. $\dfrac{\cancel{2}}{3} \times \dfrac{1}{\cancel{2}} = \dfrac{1}{3}$

17. $4 \text{ hr} - 1\dfrac{2}{3} \text{ hr} = 3\dfrac{3}{3} \text{ hr} - 1\dfrac{2}{3} \text{ hr} = 2\dfrac{1}{3} \text{ hours}$

18. $0.2 \text{ gal/mi} \times 56.4 \text{ mi} = 11.28 \text{ gal}$

19. $432 \text{ km} \times 0.6 \text{ mi/km} = 259.2 \text{ mi}$

20. 4,000 meters
    approximately 4,000 yards

Lesson Practice 11A

1. done
2. done
3. $30\% = 0.30$
4. $85\% = 0.85$
5. $10\% = 0.1$
6. $9\% = 0.09$
7. done
8. $5\% = \dfrac{5}{100} = \dfrac{1}{20}$
9. $75\% = \dfrac{75}{100} = \dfrac{3}{4}$
10. $\dfrac{1}{1} = 1.00 = 100\%$
11. $\dfrac{1}{2} = 0.5 = 50\%$
12. $\dfrac{1}{4} = 0.25 = 25\%$
13. $\dfrac{3}{4} = 0.75 = 75\%$
14. $\dfrac{1}{5} = 0.2 = 20\%$
15. $\$3.00 \times 0.06 = \$0.18$
16. $\$7.60 \times 0.15 = \$1.14$
17. $\$7.60 \times 0.50 = \$3.80 \text{ off}$
    $\$7.60 - \$3.80 = \$3.80 \text{ sale price}$
18. $\$3.80 \times 0.05 = \$0.19 \text{ tax}$
    $\$3.80 + \$0.19 = \$3.99 \text{ total}$

Lesson Practice 11B

1. $38\% = 0.38$
2. $1\% = 0.01$
3. $95\% = 0.95$
4. $71\% = 0.71$
5. $15\% = 0.15$
6. $3\% = 0.03$

7. $10\% = \dfrac{10}{100} = \dfrac{1}{10}$
8. $23\% = \dfrac{23}{100}$
9. $60\% = \dfrac{60}{100} = \dfrac{3}{5}$
10. $\dfrac{1}{4} = 0.25 = 25\%$
11. $\dfrac{3}{4} = 0.75 = 75\%$
12. $\dfrac{1}{5} = 0.2 = 20\%$
13. $\dfrac{1}{1} = 1.00 = 100\%$
14. $\dfrac{1}{2} = 0.5 = 50\%$
15. $\$7.00 \times 0.05 = \$0.35$
16. $\$10.50 \times 0.16 = \$1.68$
17. $\$12.50 \times 0.20 = \$2.50 \text{ off}$
    $\$12.50 - \$2.50 = \$10.00 \text{ sale price}$
    $\$10.00 \times 0.09 = \$0.90 \text{ tax}$
    $\$10.00 + \$0.90 = \$10.90 \text{ total}$
18. $0.25 \times 80 = 20 \text{ happy people}$

Lesson Practice 11C

1. $47\% = 0.47$
2. $5\% = 0.05$
3. $69\% = 0.69$
4. $18\% = 0.18$
5. $32\% = 0.32$
6. $2\% = 0.02$
7. $15\% = \dfrac{15}{100} = \dfrac{3}{20}$
8. $17\% = \dfrac{17}{100}$
9. $50\% = \dfrac{50}{100} = \dfrac{1}{2}$
10. $\dfrac{1}{2} = 0.5 = 50\%$
11. $\dfrac{1}{5} = 0.2 = 20\%$
12. $\dfrac{3}{4} = 0.75 = 75\%$

13. $\dfrac{1}{4} = 0.25 = 25\%$

14. $\dfrac{1}{1} = 1.00 = 100\%$

15. done

16. $\$45.00 \times 0.30 = \$13.50$ off
$\$45.00 - \$13.50 = \$31.50$ sale price
$\$31.50 \times 0.08 = \$2.52$ tax
$\$31.50 + \$2.52 = \$34.02$ total

17. $\$18.50 + \$9.50 = \$28.00$ total meals
$\$28.00 \times 0.2 = \$5.60$ tip
$\$28.00 + \$5.60 = \$33.60$ total meals and tip

18. $0.50 \times 20 = 10$ people own sleds
or $\dfrac{1}{2} \times 20 = 10$ people own sleds

Systematic Review 11D

1. done

2. $88\% = 0.88 = \dfrac{88}{100} = \dfrac{22}{25}$

3. $100\% = 1 = \dfrac{100}{100} = 1$

4. $25\% = 0.25 = \dfrac{25}{100} = \dfrac{1}{4}$

5. 50,000 cl

6. 1,200 dm

7. $200 \div 2 = 100$

8. $84 \div 4 = 21$
$21 \times 3 = 63$

9. $99 \div 3 = 33$

10. 64

11. 1

12. 1,000

13. done

14. $5\dfrac{5}{8} \times 1\dfrac{2}{5} = \dfrac{\overset{9}{\cancel{45}}}{8} \times \dfrac{7}{\cancel{5}} = \dfrac{63}{8} = 7\dfrac{7}{8}$

15. $50 \times 0.96 = 48$ questions right
$50 - 48 = 2$ questions wrong

16. $\$120.00 \times 0.25 = \$30.00$ off
$\$120.00 - \$30.00 = \$90.00$ sale price
$\$90.00 \times 0.05 = \$4.50$ tax
$\$90.00 + \$4.50 = \$94.50$ total

17. 1,000 pieces
approximately 1 yard

18. $2\dfrac{2}{3} \times 4\dfrac{1}{2} = \dfrac{\overset{4}{\cancel{8}}}{\cancel{3}} \times \dfrac{9^{3}}{\cancel{2}} = \dfrac{12}{1} = 12$ miles

Systematic Review 11E

1. $30\% = 0.3 = \dfrac{30}{100} = \dfrac{3}{10}$

2. $72\% = 0.72 = \dfrac{72}{100} = \dfrac{18}{25}$

3. $50\% = 0.5 = \dfrac{50}{100} = \dfrac{1}{2}$

4. $20\% = 0.2 = \dfrac{20}{100} = \dfrac{1}{5}$

5. 50 cl

6. 18,000 ml

7. $120 \div 3 = 40$

8. $72 \div 9 = 8$
$8 \times 2 = 16$

9. $600 \div 6 = 100$
$100 \times 5 = 500$

10. 2.448

11. 0.36

12. 0.31

13. 1.26

14. $4\dfrac{2}{7} \times 3\dfrac{1}{5} = \dfrac{\overset{6}{\cancel{30}}}{7} \times \dfrac{16}{\cancel{5}} = \dfrac{96}{7} = 13\dfrac{5}{7}$

15. $9\dfrac{1}{5} \times 5\dfrac{4}{5} = \dfrac{46}{5} \times \dfrac{29}{5} = \dfrac{1,334}{25} = 53\dfrac{9}{25}$

16. 45% of $20 = 0.45 \times 20 = 9$ boys

17. $\$85.00 \times 0.08 = \$6.80$
shipping and handling
$\$85.00 + \$6.80 = \$91.80$ total
$\$91.80 - \$57.00 = \$34.80$ still needed

18. 100 steps

19. $\$9.95 \times 5 = \$49.75$

20. 0.52 lb $+ 0.75$ lb $= 1.27$ lb
1.27 lb $\times 0.5 = 0.635$ lb

Systematic Review 11F

1. $55\% = 0.55 = \dfrac{55}{100} = \dfrac{11}{20}$

2. $40\% = 0.4 = \dfrac{40}{100} = \dfrac{2}{5}$

3. $75\% = 0.75 = \dfrac{75}{100} = \dfrac{3}{4}$

4. $24\% = 0.24 = \dfrac{24}{100} = \dfrac{6}{25}$

5. 2,200 dg

6. 100 dag

7. $6\dfrac{2}{3} + 1\dfrac{7}{9} = 6\dfrac{18}{27} + 1\dfrac{21}{27} =$
$7\dfrac{39}{27} = 8\dfrac{12}{27} = 8\dfrac{4}{9}$

8. $7\dfrac{3}{10} + 9\dfrac{4}{5} = 7\dfrac{15}{50} + 9\dfrac{40}{50} =$
$16\dfrac{55}{50} = 17\dfrac{5}{50} = 17\dfrac{1}{10}$

9. $3\dfrac{2}{7} - 1\dfrac{2}{3} = 3\dfrac{6}{21} - 1\dfrac{14}{21} =$
$2\dfrac{27}{21} - 1\dfrac{14}{21} = 1\dfrac{13}{21}$

10. 0.8547

11. 4.32

12. 1.77

13. 3.33

14. $1\dfrac{2}{3} \times 4\dfrac{3}{5} = \dfrac{\cancel{5}}{3} \times \dfrac{23}{\cancel{5}} = \dfrac{23}{3} = 7\dfrac{2}{3}$

15. $7\dfrac{3}{5} \times 1\dfrac{1}{2} = \dfrac{\overset{19}{\cancel{38}}}{5} \times \dfrac{3}{\cancel{2}} = \dfrac{57}{5} = 11\dfrac{2}{5}$

16. 85% of $20 = 0.85 \times 20 = 17$ games won

17. $\$1.99 \times 0.15 = \$0.2985$
(rounded to $0.30) off
$\$1.99 - \$0.30 = \$1.69$ for one fish
$\$1.69 \times 3 = \$5.07$ for three fish

18. 40 km × 0.6 mi/km = 24 mi

19. 8 L × 1.06 qt/L = 8.48 qt

20. $6\dfrac{2}{3} \times \dfrac{3}{5} = \dfrac{\overset{4}{\cancel{20}}}{\cancel{3}} \times \dfrac{\cancel{3}}{\cancel{5}} = \dfrac{4}{1} =$
$\dfrac{4}{1} = 4$ bales spoiled

Lesson Practice 12A

1. done

2. $9 = \dfrac{900}{100} = 900\%$

3. $5 = \dfrac{500}{100} = 500\%$

4. done

5. $\dfrac{1}{4} = \dfrac{25}{100} = 25\%$

6. $\dfrac{3}{4} = \dfrac{75}{100} = 75\%$

7. $\dfrac{1}{5} = \dfrac{20}{100} = 20\%$

8. $\dfrac{2}{5} = \dfrac{40}{100} = 40\%$

9. $\dfrac{3}{5} = \dfrac{60}{100} = 60\%$

10. $1\dfrac{1}{4} = \dfrac{100}{100} + \dfrac{25}{100} = \dfrac{125}{100} = 125\%$

11. $2\dfrac{1}{2} = \dfrac{200}{100} + \dfrac{50}{100} = \dfrac{250}{100} = 250\%$

12. $3\dfrac{3}{4} = \dfrac{300}{100} + \dfrac{75}{100} = \dfrac{375}{100} = 375\%$

13. $5\dfrac{1}{5} = \dfrac{500}{100} + \dfrac{20}{100} = \dfrac{520}{100} = 520\%$

14. done

15. $250\% = 2.5$

16. $520\% = 5.2$

17. $100\% + 15\% + 6\% = 121\% = 1.21$
$1.21 \times \$18.90 = \$22.869$
(rounds to $22.87)

18. $400\% = 4$
$4 \times \$0.45 = \$1.80$

Lesson Practice 12B

1. $8 = \dfrac{800}{100} = 800\%$

2. $6 = \dfrac{600}{100} = 600\%$

3. $2 = \dfrac{200}{100} = 200\%$

4. $\dfrac{2}{5} = \dfrac{40}{100} = 40\%$

5. $\dfrac{1}{5} = \dfrac{20}{100} = 20\%$

6. $\dfrac{4}{5} = \dfrac{80}{100} = 80\%$

7. $\dfrac{1}{10} = \dfrac{10}{100} = 10\%$

8. $\dfrac{9}{10} = \dfrac{90}{100} = 90\%$

9. $\dfrac{1}{4} = \dfrac{25}{100} = 25\%$

10. $3\dfrac{1}{10} = \dfrac{300}{100} + \dfrac{10}{100} = \dfrac{310}{100} = 310\%$

11. $6\dfrac{3}{4} = \dfrac{600}{100} + \dfrac{75}{100} = \dfrac{675}{100} = 675\%$

12. $5\dfrac{1}{2} = \dfrac{500}{100} + \dfrac{50}{100} = \dfrac{550}{100} = 550\%$

13. $2\dfrac{3}{5} = \dfrac{200}{100} + \dfrac{60}{100} = \dfrac{260}{100} = 260\%$

14. $310\% = 3.10$

15. $675\% = 6.75$

16. $100\% = 1$

17. $100\% + 20\% + 5\% = 125\% = 1.25$
$1.25 \times \$30.00 = \$37.50$

18. $100\% + 5\% = 105\% = 1.05$
$1.05 \times \$32.95 = \$34.5975$
(rounds to 34.60)

Lesson Practice 12C

1. $3 = \dfrac{300}{100} = 300\%$

2. $7 = \dfrac{700}{100} = 700\%$

3. $1 = \dfrac{100}{100} = 100\%$

4. $\dfrac{1}{2} = \dfrac{50}{100} = 50\%$

5. $\dfrac{3}{4} = \dfrac{75}{100} = 75\%$

6. $\dfrac{3}{5} = \dfrac{60}{100} = 60\%$

7. $\dfrac{5}{10} = \dfrac{50}{100} = 50\%$

8. $\dfrac{1}{25} = \dfrac{4}{100} = 4\%$

9. $\dfrac{1}{50} = \dfrac{2}{100} = 2\%$

10. $7\dfrac{1}{4} = \dfrac{700}{100} + \dfrac{25}{100} = \dfrac{725}{100} = 725\%$

11. $4\dfrac{1}{5} = \dfrac{400}{100} + \dfrac{20}{100} = \dfrac{420}{100} = 420\%$

12. $1\dfrac{3}{4} = \dfrac{100}{100} + \dfrac{75}{100} = \dfrac{175}{100} = 175\%$

13. $3\dfrac{5}{10} = \dfrac{300}{100} + \dfrac{50}{100} = \dfrac{350}{100} = 350\%$

14. $725\% = 7.25$

15. $420\% = 4.2$

16. $850\% = 8.5$

17. $100\% + 18\% + 2\% = 120\% = 1.2$
$1.2 \times \$55.20 = \$66.24$

18. $100\% = 1$
$1 \times 20 = 20$

Systematic Review 12D

1. done

2. $6\dfrac{3}{4} = \dfrac{600}{100} + \dfrac{75}{100} = \dfrac{675}{100} = 6.75 = 675\%$

3. $25\% = 0.25 = \dfrac{25}{100} = \dfrac{1}{4}$

4. $50\% = 0.5 = \dfrac{50}{100} = \dfrac{1}{2}$

5. 600 m

6. $7,000 \text{ mm}$

7. $1,000$

8. 100

9. 10

10. $\dfrac{1}{10}$

11. $\dfrac{1}{100}$

12. $\dfrac{1}{1,000}$

13. $\dfrac{3}{4} \div \dfrac{1}{4} = \dfrac{3 \div 1}{1} = 3$

14. $\dfrac{4}{5} \div \dfrac{1}{3} = \dfrac{12}{15} \div \dfrac{5}{15} = \dfrac{12 \div 5}{1} = 2\dfrac{2}{5}$

15. $\dfrac{2}{3} \div \dfrac{1}{4} = \dfrac{8}{12} \div \dfrac{3}{12} = \dfrac{8 \div 3}{1} = \dfrac{8}{3} = 2\dfrac{2}{3}$

16. $100\% + 6\% + 5\% = 111\% = 1.11$
$1.11 \times \$36.00 = \$39.96$

17. $100\% - 25\% = 75\% = 0.75$
$0.75 \times \$39.96 = \$29.97$

18. Yes; $\left(10\% = 0.1 = \dfrac{1}{10} \right)$
$0.1 \times 100 = 10$

Systematic Review 12E

1. $2\dfrac{1}{4} = \dfrac{200}{100} + \dfrac{25}{100} = \dfrac{225}{100} =$
   $225\% = 2.25$

2. $8\dfrac{1}{5} = \dfrac{800}{100} + \dfrac{20}{100} = \dfrac{820}{100} =$
   $820\% = 8.20$

3. $42\% = 0.42 = \dfrac{42}{100} = \dfrac{21}{50}$

4. $150\% = 1.5 = \dfrac{150}{100} = 1\dfrac{50}{100} = 1\dfrac{1}{2}$

5. yard
6. inches
7. quarts
8. ounce
9. pounds
10. miles
11. inch

12. $\dfrac{7}{8} \div \dfrac{1}{3} = \dfrac{21}{24} \div \dfrac{8}{24} = \dfrac{21 \div 8}{1} =$
    $\dfrac{21}{8} = 2\dfrac{5}{8}$

13. $\dfrac{5}{6} \div \dfrac{1}{2} = \dfrac{10}{12} \div \dfrac{6}{12} = \dfrac{10 \div 6}{1} =$
    $\dfrac{10}{6} = 1\dfrac{4}{6} = 1\dfrac{2}{3}$

14. $\dfrac{2}{3} \div \dfrac{1}{4} = \dfrac{8}{12} \div \dfrac{3}{12} = \dfrac{8 \div 3}{1} =$
    $\dfrac{8}{3} = 2\dfrac{2}{3}$

15. $\dfrac{5}{8} \div \dfrac{1}{16} = \dfrac{80}{128} \div \dfrac{8}{128} =$
    $\dfrac{80 \div 8}{1} = \dfrac{10}{1} = 10$ pies

16. $\$235.00 \times 0.05 = \$11.75$ tax
    $\$235.00 + \$11.75 = \$246.75$

17. $100\% - 40\% = 60\% = 0.6$
    $0.6 \times \$12.4 = \$7.44$

18. 68×2.5 cm $= 170$ cm

19. 170 cm $\times 10$ mm/cm $= 1{,}700$ mm

20. Yes; $\left(50\% = \dfrac{1}{2}\right)$

    $\dfrac{1}{2} \times \dfrac{\$40.00}{1} = \dfrac{\$40.00}{2} = \$20.00$

    or

    $0.5 \times \$40.00 = \$20.00$

Systematic Review 12F

1. $7\dfrac{3}{4} = \dfrac{700}{100} + \dfrac{75}{100} = \dfrac{775}{100} =$
   $775\% = 7.75$

2. $5\dfrac{1}{10} = \dfrac{500}{100} + \dfrac{10}{100} = \dfrac{510}{100} =$
   $510\% = 5.10$

3. $9\% = 0.09 = \dfrac{9}{100}$

4. $100\% = 1 = \dfrac{100}{100} = 1$

5. 3 feet or 1 yard
6. 0.4 inch
7. 2.5 cm; 2.54 cm
8. 28 grams
9. 2.2 lb
10. 0.6 miles
11. 1.06 qt

12. $\dfrac{6}{7} \div \dfrac{4}{7} = \dfrac{6 \div 4}{1} = \dfrac{6}{4} = 1\dfrac{2}{4} = 1\dfrac{1}{2}$

13. $\dfrac{3}{5} \div \dfrac{1}{3} = \dfrac{9}{15} \div \dfrac{5}{15} = \dfrac{9 \div 5}{1} = \dfrac{9}{5} = 1\dfrac{4}{5}$

14. $\dfrac{5}{8} \div \dfrac{2}{3} = \dfrac{15}{24} \div \dfrac{16}{24} = \dfrac{15 \div 16}{1} = \dfrac{15}{16}$

15. $\dfrac{4}{5} \div \dfrac{2}{5} = \dfrac{4 \div 2}{1} = \dfrac{4}{2} = 2$ friends

16. $20 \times 3 = 60$ balloons

17. $100\% + 20\% + 4\% = 124\% = 1.24$
    $1.24 \times \$15.00 = \$18.60$

18. 20 oz $\times 28$ g/oz $= 560$ g

19. 560 g $\times 1{,}000$ mg/g $= 560{,}000$ mg

20. $100\% - 20\% = 80\% = 0.8$
    $0.8 \times \$355.00 = \$284.00$

Lesson Practice 13A

1. brown
2. $0.5 \times 20 = 10$ people have brown hair
3. 5 $\left(\text{There are the same number with blond and red hair.}\right)$
4. vanilla
5. strawberry
6. $300 \times 0.3 = 90$ chocolate cones

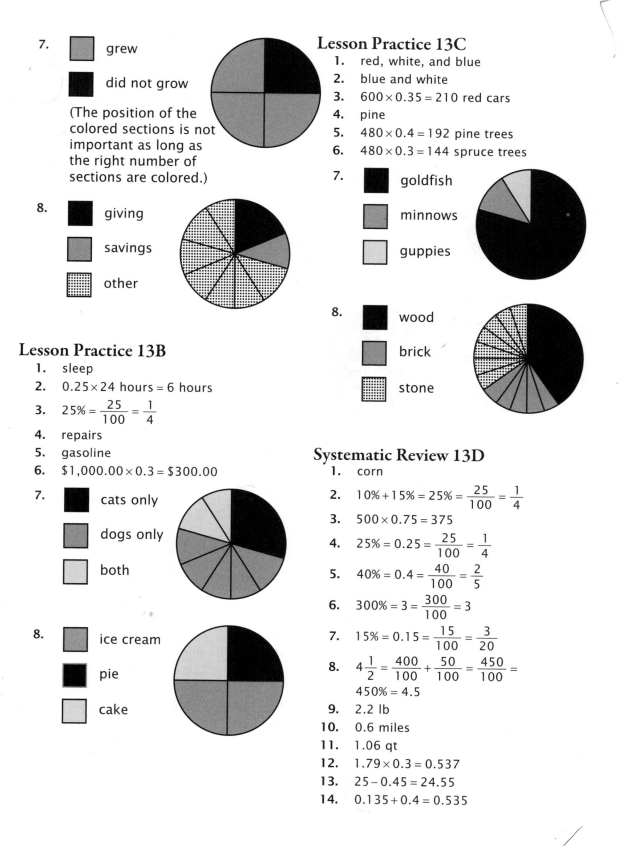

7. ■ grew

■ did not grow

(The position of the colored sections is not important as long as the right number of sections are colored.)

8. ■ giving

■ savings

▦ other

Lesson Practice 13B

1. sleep
2. $0.25 \times 24 \text{ hours} = 6 \text{ hours}$
3. $25\% = \dfrac{25}{100} = \dfrac{1}{4}$
4. repairs
5. gasoline
6. $\$1,000.00 \times 0.3 = \$300.00$
7. ■ cats only

■ dogs only

□ both
8. ■ ice cream

■ pie

□ cake

Lesson Practice 13C

1. red, white, and blue
2. blue and white
3. $600 \times 0.35 = 210 \text{ red cars}$
4. pine
5. $480 \times 0.4 = 192 \text{ pine trees}$
6. $480 \times 0.3 = 144 \text{ spruce trees}$
7. ■ goldfish

■ minnows

□ guppies
8. ■ wood

■ brick

▦ stone

Systematic Review 13D

1. corn
2. $10\% + 15\% = 25\% = \dfrac{25}{100} = \dfrac{1}{4}$
3. $500 \times 0.75 = 375$
4. $25\% = 0.25 = \dfrac{25}{100} = \dfrac{1}{4}$
5. $40\% = 0.4 = \dfrac{40}{100} = \dfrac{2}{5}$
6. $300\% = 3 = \dfrac{300}{100} = 3$
7. $15\% = 0.15 = \dfrac{15}{100} = \dfrac{3}{20}$
8. $4\dfrac{1}{2} = \dfrac{400}{100} + \dfrac{50}{100} = \dfrac{450}{100} =$
   $450\% = 4.5$
9. 2.2 lb
10. 0.6 miles
11. 1.06 qt
12. $1.79 \times 0.3 = 0.537$
13. $25 - 0.45 = 24.55$
14. $0.135 + 0.4 = 0.535$

15. $4\frac{2}{3} \div 1\frac{1}{5} = \frac{14}{3} \div \frac{6}{5} = \frac{70}{15} \div \frac{18}{15} =$

$\frac{70 \div 18}{1} = \frac{70}{18} = \frac{35}{9} = 3\frac{8}{9}$

16. $4\frac{1}{8} \div 2\frac{2}{3} = \frac{33}{8} \div \frac{8}{3} = \frac{99}{24} \div \frac{64}{24} =$

$\frac{99 \div 64}{1} = \frac{99}{64} = 1\frac{35}{64}$

17. $7\frac{5}{10} \div 1\frac{1}{2} = \frac{75}{10} \div \frac{3}{2} = \frac{150}{20} \div \frac{30}{20} =$

$\frac{150 \div 30}{1} = \frac{150}{30} = 5$ bags

18. $100\% + 7\% = 107\% = 1.07$
$\$35.00 \times 1.07 = \$37.45$

Systematic Review 13E

1. biking / running / swimming

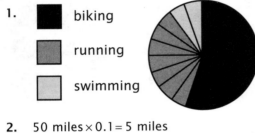

2. 50 miles $\times 0.1 = 5$ miles

3. $50\% = 0.5 = \frac{50}{100} = \frac{1}{2}$

4. $20\% = 0.2 = \frac{20}{100} = \frac{1}{5}$

5. $5\% = 0.05 = \frac{5}{100} = \frac{1}{20}$

6. $700\% = 7 = \frac{700}{100} = 7$

7. $6\frac{1}{4} = \frac{600}{100} + \frac{25}{100} =$

$\frac{625}{100} = 625\% = 6.25$

8. 0.4 inches

9. 2.5 cm (or exactly 2.54 cm)

10. 28 grams

11. $2.45 \times 0.09 = .2205$

12. $0.7 - 0.15 = 0.55$

13. $1.58 + 7.6 = 9.18$

14. $3\frac{1}{6} \div 1\frac{1}{5} = \frac{19}{6} \div \frac{6}{5} = \frac{95}{30} \div \frac{36}{30} =$

$= \frac{95 \div 36}{1} = \frac{95}{36} = 2\frac{23}{36}$

15. $1\frac{1}{4} \div 1\frac{1}{8} = \frac{5}{4} \div \frac{9}{8} = \frac{40}{32} \div \frac{36}{32}$

$= \frac{40 \div 36}{1} = \frac{40}{36} = \frac{10}{9} = 1\frac{1}{9}$

16. $2\frac{1}{6} \div 1\frac{4}{5} = \frac{13}{6} \div \frac{9}{5} = \frac{65}{30} \div \frac{54}{30}$

$= \frac{65 \div 54}{1} = \frac{65}{54} = 1\frac{11}{54}$

17. $6\frac{3}{4} \div 1\frac{1}{4} = \frac{27}{4} \div \frac{5}{4} = \frac{27 \div 5}{1}$

$= \frac{27}{5} = 5\frac{2}{5}$

5 people can be served. There will be pizza left over because she can't serve pizza to a part of a person.

18. $100\% + 10\% = 110\% = 1.1$
$\$1.30 \times 1.1 = \$1.43$

19. 55.6 mi/hr $\times 7$ hr $= 389.2$ miles

20. 2 kg $\times 2.2$ lb/kg $= 4.4$ lb

Systematic Review 13F

1. clouds

2. $30 \times 0.2 = 6$ days

3. $30 \times 0.1 = 3$ days

4. $75\% = 0.75 = \frac{75}{100} = \frac{3}{4}$

5. $2\% = 0.02 = \frac{2}{100} = \frac{1}{50}$

6. $500\% = 5 = \frac{500}{100} = 5$

7. $30\% = 0.3 = \frac{30}{100} = \frac{3}{10}$

8. $3\frac{1}{5} = \frac{300}{100} + \frac{20}{100} = \frac{320}{100} = 320\% = 3.2$

9. 6,000 g

10. 25,000 ml

11. $0.17 \times 0.58 = 0.0986$

12. $1.75 - 0.9 = 0.85$

13. $3.68 + 0.061 = 3.741$

14. $3\frac{1}{2} \div 2\frac{2}{5} = \frac{7}{2} \div \frac{12}{5} = \frac{35}{10} \div \frac{24}{10} =$

$\frac{35 \div 24}{1} = \frac{35}{24} = 1\frac{11}{24}$

15. $5\frac{1}{2} \div 2\frac{2}{3} = \frac{11}{2} \div \frac{8}{3} = \frac{33}{6} \div \frac{16}{6} =$

$\frac{33 \div 16}{1} = \frac{33}{16} = 2\frac{1}{16}$

16. $3\frac{1}{5} \div 1\frac{3}{4} = \frac{16}{5} \div \frac{7}{4} = \frac{64}{20} \div \frac{35}{20} =$

$\frac{64 \div 35}{1} = \frac{64}{35} = 1\frac{29}{35}$

17. 200 cm

200 cm × 0.4 in/cm = 80 in

18. 60 × 1.2 = 72 donuts

19. 1.8 cups/hr × 8 hr = 14.4 cups

20. 50 km × 0.6 mi/km = 30 mi

Lesson Practice 14A

1. done

2. done

3. Estimate:

```
  20.
×  .1
  2.0
```

```
  22.163        22 163   3 places
×   .11       ×     11   2 places
  .22163          22163
 2.2163          2 2163
 2.43793        2.43793  5 places
```

4. Estimate:

```
  8
× 2
 16
```

```
  8.246          8246   3 places
× 2.159       × 2 159   3 places
 .074214        742 14
 .41230        4 1230
 .8246          8246
16.492        16 492
17.803114     17.803114  6 places
```

5. Estimate:

```
 100.
×  .6
  60.
```

```
  135.004        1 35004  3 places
×    .58       ×      58  2 places
  10.80032       10 80032
  67.5020        67 5020
  78.30232       78.30232  5 places
```

6. Estimate:

```
  40.
×  .6
  24.
```

```
  41.685         4 1685   3 places
×   .555       ×    555   3 places
  .208425         208425
 2.08425         2 08425
20.8425         208425
23.135175       23.135175  6 places
```

7. 9.625 mi/hr × 0.25 hr = 2.40625 mi

8. 12.75 × $1.539 = $19.62225
(rounded to $19.62)

Lesson Practice 14B

1. Estimate:

```
  50.
× .005
 .25
```

```
  50.345          50345  3 places
×   .005       ×    005  3 places
 .251725         .251725  6 places
```

2. Estimate:

```
 31
× 2
 62
```

```
  31.489          3 1489  3 places
×  2.23        ×     223  2 places
  .94467          94467
 6.2978          6 2978
62.978          62 978
70.22047        70.22047  5 places
```

3. Estimate:

```
    .9
 ×.009
  .0081

   .876            876   3 places
 ×.009       ×        9   3 places
 (.007884)      .007884   6 places
```

4. Estimate :

```
   6
  ×1
   6

   6.218          6218    3 places
 ×1.064      ×     1064    3 places
  .024872         24872
  .37308          37308
  6.218           6 218
 (6.615952)       6.615952  6 places
```

5. Estimate:

```
   500.
 ×    .2
  100

  450.132         450 132   3 places
 ×   .222     ×       222   3 places
  .900264          900264
  9.00264          9 00264
  90.0264          90 0264
 (99.929304)       99.929304  6 places
```

6. Estimate:

```
    2
  × .8
   1.6

   1.539          1539     3 places
 ×  .82       ×     82     2 places
  .03078           3078
  1.2312           1 23 12
 (1.26198)         1.26 198  5 places
```

7. 39.37 in/m × 0.001 m = 0.03937 in
8. $1.279 × 1.30 = $1.6627
   (rounded to $1.663)

Lesson Practice 14C

1. Estimate:

```
   60
 ×   .03
  1.80

  61.198           6 1198   3 places
 ×   .03       ×        3   2 places
  1.83594          1.83594  5 places
```

2. Estimate:

```
   40
 ×  1
   40

  42.345          42345    3 places
 ×  1.16      ×     116    2 places
  2.54070          2 54070
  4.2345           4 2345
  42.345           42.345
  49.12020         49.12020  5 places
```

3. Estimate:

```
   .5
 ×.8
   .40

   .511           5 11     3 places
 ×.8          ×      8     1 place
  .4088            .4088    4 places
```

4. Estimate:

```
    8
  × 3
   24

   7.639          7639     3 places
 × 2.804      ×    2804     3 places
  .030556          30556
  6.1112           6 11120
  15.278           15 278
  21.419756        21. 4 19756  6 places
```

5. Estimate:

$$\begin{array}{r} 20 \\ \times\ .1 \\ \hline 2 \end{array}$$

$$\begin{array}{r} 20.352 \\ \times\ .136 \\ \hline .122112 \\ .61056 \\ 2.0352 \\ \hline 2.767872 \end{array}$$

$$\begin{array}{r} 20352 \quad \text{3 places} \\ \times\ \quad 136 \quad \text{3 places} \\ \hline 122112 \\ 61056 \\ 2\,0352 \\ \hline 2.767872 \quad \text{6 places} \end{array}$$

6. Estimate:

$$\begin{array}{r} 9 \\ \times\ .7 \\ \hline 6.3 \end{array}$$

$$\begin{array}{r} 9.293 \\ \times\ .71 \\ \hline .09293 \\ 6.5051 \\ \hline 6.59803 \end{array}$$

$$\begin{array}{r} 9293 \quad \text{3 places} \\ \times\ \quad 71 \quad \text{2 places} \\ \hline 9293 \\ 6\,5051 \\ \hline 6.59803 \quad \text{5 places} \end{array}$$

$$\begin{array}{r} 9293 \quad \text{3 places} \\ \times 71 \quad +\text{2 places} \\ \hline 9293 \\ 65051 \\ \hline 6.59803 \quad \text{5 places} \end{array}$$

7. $1.06 \text{ qt/L} \times 0.001 \text{ L} = 0.00106 \text{ qt}$

8. $1.667 \text{ gal/min} \times 60 \text{ min} = 100.02 \text{ gal}$

Systematic Review 14D

1. Estimate:

$$\begin{array}{r} 2 \\ \times\ .6 \\ \hline 1.2 \end{array}$$

$$\begin{array}{r} 1724 \quad \text{3 places} \\ \times\ \quad 6 \quad \text{1 place} \\ \hline 1.0344 \quad \text{4 places} \end{array}$$

2. Estimate:

$$\begin{array}{r} 3 \\ \times\ .4 \\ \hline 1.2 \end{array}$$

$$\begin{array}{r} 3465 \quad \text{3 places} \\ \times\ \quad 403 \quad \text{3 places} \\ \hline 10395 \\ 13860 \\ \hline 1.396395 \quad \text{6 places} \end{array}$$

3. $50\% = 0.5 = \dfrac{50}{100} = \dfrac{1}{2}$

4. $8\% = 0.08 = \dfrac{8}{100} = \dfrac{2}{25}$

5. $600\% = 6 = \dfrac{600}{100} = 6$

6. $20\% = 0.2 - \dfrac{20}{100} = \dfrac{1}{5}$

7. $5\dfrac{1}{4} = \dfrac{500}{100} + \dfrac{25}{100} = \dfrac{525}{100} = 525\% = 5.25$

8. $70,000 \text{ cm}$

9. 100 mg

10. $6\dfrac{1}{8} + 3\dfrac{1}{2} = 6\dfrac{2}{16} + 3\dfrac{8}{16} = 9\dfrac{10}{16} = 9\dfrac{5}{8}$

11. $10\dfrac{2}{5} - 4\dfrac{5}{8} = 10\dfrac{16}{40} - 4\dfrac{25}{40} =$

$9\dfrac{56}{40} - 4\dfrac{25}{40} = 5\dfrac{31}{40}$

12. $\dfrac{\cancel{7}}{3} \times \dfrac{\cancel{6}^{2}}{\cancel{7}} = \dfrac{2}{1} = 2$

13. $\dfrac{3}{5} \times \dfrac{2}{1} = \dfrac{6}{5} = 1\dfrac{1}{5}$

14. $\dfrac{7}{\cancel{8}_2} \times \dfrac{\cancel{4}}{3} = \dfrac{7}{6} = 1\dfrac{1}{6}$

15. $\dfrac{3}{8} \times \dfrac{5}{1} = \dfrac{15}{8} = 1\dfrac{7}{8}$

16. $\dfrac{10}{16} \div \dfrac{1}{8} = \dfrac{10}{\cancel{16}_2} \times \dfrac{\cancel{8}}{1} = \dfrac{10}{2} = 5 \text{ pieces}$

17. $\$15.50 \times 1.06 = \$16.43$

18. $0.6 \text{ mi/km} \times 0.001 \text{ km} = 0.0006 \text{ mi}$

Systematic Review 14E

1. Estimate:

$$\begin{array}{r} 2 \\ \times\ .007 \\ \hline .014 \end{array}$$

$$\begin{array}{r} 2192 \quad \text{3 places} \\ \times\ \quad 7 \quad \text{3 places} \\ \hline .015344 \quad \text{6 places} \end{array}$$

2. Estimate:

$$\begin{array}{r} 10 \\ \times\ .8 \\ \hline 8 \end{array}$$

$$\begin{array}{r} 13543 \quad \text{3 places} \\ \times\ \quad 75 \quad \text{2 places} \\ \hline 67715 \\ 94801 \\ \hline 10.15725 \quad \text{5 places} \end{array}$$

3. summer

4. spring

5. $850 \times 0.10 = 85$

6. $25\% = 0.25 - \dfrac{25}{100} = \dfrac{1}{4}$

7. $100\% = 1 = \dfrac{100}{100} = 1$

8. $250\% = 2.5 = \dfrac{250}{100} = 2\dfrac{1}{2}$

9. $16\% = 0.16 = \dfrac{16}{100} = \dfrac{4}{25}$

10. $1\dfrac{3}{4} = \dfrac{100}{100} + \dfrac{75}{100} = \dfrac{175}{100} = 175\% = 1.75$

11. 12,000 L

12. 300 cm

13. $\dfrac{5}{\cancel{9}_3} \times \dfrac{\cancel{3}}{2} = \dfrac{5}{6}$

14. $\dfrac{9}{\cancel{2}} \times \dfrac{\cancel{5}}{1} = \dfrac{9}{2} = 4\dfrac{1}{2}$

15. $\dfrac{1}{3} \times \dfrac{4}{3} = \dfrac{4}{9}$

16. $8\ hr \times 0.08 = 0.64$ hours

17. $15.25 \times \$1.549 =$
$\$23.62225$ (rounded \$23.62)

18. $1.56\ lb + 2.13\ lb + 1.5\ lb + 0.66\ lb = 5.85\ lb$

19. $5.85\ lb \times 16\ oz/lb = 93.6\ oz$
$93.6\ oz \times 28\ g/oz = 2{,}620.8\ g$

20. $0.25 \times 12 = 3$ oranges to her mother
$\dfrac{1}{4} \times \dfrac{12}{1} = \dfrac{12}{4} = 3$ oranges to her sister
$12 - 6 = 6$ left
$6 \div 12 = 0.5 = 50\%$ left

Systematic Review 14F

1. Estimate:

```
  .6              567   3 places
 ×.1          ×   148   3 places
 ----             -----
  .06             4536
                  2268
                   567
                 --------
                  .083916  6 places
```

2. Estimate:

```
   2            1905   3 places
  ×1         ×  1321   3 places
 ----          -----
   2            1905
               3810
               5715
              1905
             ----------
              2.516505  6 places
```

3.

■ oak
▨ maple
▢ hemlock

4. $600 \times 0.05 = 30$ trees are hemlock

5. $75\% = 0.75 = \dfrac{75}{100} = \dfrac{3}{4}$

6. $45\% = 0.45 - \dfrac{45}{100} = \dfrac{9}{20}$

7. $125\% = 1.25 = \dfrac{125}{100} = 1\dfrac{1}{4}$

8. $1\% = 0.01 = \dfrac{1}{100}$

9. $7\dfrac{1}{5} = \dfrac{700}{100} + \dfrac{20}{100} = \dfrac{720}{100} = 720\% = 7.2$

10. 35,000 mg

11. 1,100 dam

12. 500 cm

13. $\dfrac{2}{5} \times \dfrac{8}{3} = \dfrac{16}{15} = 1\dfrac{1}{15}$

14. $\dfrac{\cancel{4}^2}{5} \times \dfrac{3}{\cancel{2}} = \dfrac{6}{5} = 1\dfrac{1}{5}$

15. $\dfrac{\cancel{3}}{\cancel{4}} \times \dfrac{\cancel{4}}{\cancel{3}} = 1$

16. $900 \times 0.15 = 135$ yellow roses

17. $\$9.00 \times 1.10 = \$9.90$

18. $90\ cm \times 0.4\ in/cm = 36\ in$

19. $5.725\ tons \times 0.1 = 0.5725\ ton$

20. $0.5725\ ton \times 2{,}000\ lb/ton = 1{,}145\ lb$

Lesson Practice 15A

1. done

2. done

3. $750\ dg \times \dfrac{1\ g}{10\ dg} \times \dfrac{1\ dag}{10\ g} = 7.5\ dag$
or two places larger
so 750 dg = 7.5 dag

4. 0.014 km

5. 0.4 km

6. 5.3 L

7. 1.3 cg

8. 0.48 dam
9. 0.07 hl
10. 0.000609 kg
11. 0.5 cm
12. 1.2 dal

Lesson Practice 15B

1. 30 L
2. 5 m
3. 10 kl
4. 0.25 km
5. 0.18 g
6. 0.45 kl
7. 0.5 km
8. 80 g
9. 8.3 cg
10. 0.008 kl
11. 1.02 m
12. 0.01 km

Lesson Practice 15C

1. 0.015 dam
2. 8 g
3. 1 hm
4. 0.25 km
5. 30 g
6. 0.034 dag
7. 0.99 m
8. 7.3 dam
9. 15 kl
10. 50 m
11. 45.638 g
12. 300 kg

Systematic Review 15D

1. 0.45 g
2. 60 cm
3. 100 mg

4. Estimate:

$$\begin{array}{r} .6 \\ \times .5 \\ \hline .30 \end{array}$$

$$\begin{array}{r} 6\,2\,1 \quad \text{3 places} \\ \times \quad 5\,4\,3 \quad \text{3 places} \\ \hline 1\,8\,6\,3 \\ 2\,4\,8\,4 \\ 3\,1\,0\,5 \\ \hline .337203 \quad \text{6 places} \end{array}$$

5. Estimate:

$$\begin{array}{r} 3 \\ \times 2 \\ \hline 6 \end{array}$$

$$\begin{array}{r} 2\,7\,2\,8 \quad \text{3 places} \\ \times \quad 1\,6\,9 \quad \text{2 places} \\ \hline 2\,4\,5\,5\,2 \\ 1\,6\,3\,6\,8 \\ 2\,7\,2\,8 \\ \hline 4.61032 \quad \text{5 places} \end{array}$$

6. $11\% = 0.11 = \dfrac{11}{100}$

7. $5\% = 0.05 = \dfrac{5}{100} = \dfrac{1}{20}$

8. $100\% = 1 = \dfrac{100}{100} = 1$

9. $25\% = 0.25 = \dfrac{25}{100} = \dfrac{1}{4}$

10. $18 \div 2 = 9$

$9 \times 1 = 9$

11. $69 \div 3 = 23$

$23 \times 2 = 46$

12. $248 \div 8 = 31$

$31 \times 5 = 155$

13. $4 \div 4 = 1$

14. $7\dfrac{7}{8} \div 1\dfrac{1}{2} = \dfrac{63}{8} \div \dfrac{3}{2} =$

$\dfrac{\overset{21}{\cancel{63}}}{\underset{4}{\cancel{8}}} \times \dfrac{\cancel{2}}{\cancel{3}} = \dfrac{21}{4} = 5\dfrac{1}{4}$

15. $3\dfrac{2}{7} \div 2\dfrac{1}{3} = \dfrac{23}{7} \div \dfrac{7}{3} =$

$\dfrac{23}{7} \times \dfrac{3}{7} = \dfrac{69}{49} = 1\dfrac{20}{49}$

16. $24 \text{ strides} \times \dfrac{1 \text{ m}}{1 \text{ stride}} \times \dfrac{1 \text{ dam}}{10 \text{ m}} = 2.4 \text{ dam}$

17. $105 \text{ kg} \times 2.2 \text{ lb/kg} = 231 \text{ lb}$

18. $\$1.679 \times 2 = \$3.358$

Systematic Review 15E

1. 0.39 m
2. 3,800 g
3. 0.087 kl
4. Estimate:

$$
\begin{array}{r}
50 \\
\times\ .6 \\
\hline
30
\end{array}
$$

$$
\begin{array}{r}
48003 \quad \text{3 places} \\
\times \qquad 61 \quad \text{2 places} \\
\hline
48003 \\
28\ 80\ 18 \\
\hline
29.28183 \quad \text{5 places}
\end{array}
$$

5. Estimate:

$$
\begin{array}{r}
20 \\
\times\ 30 \\
\hline
600
\end{array}
$$

$$
\begin{array}{r}
2277 \quad \text{2 places} \\
\times \quad 3154 \quad \text{2 places} \\
\hline
9108 \\
11385 \\
2277 \\
6831 \\
\hline
718.1658 \quad \text{4 places}
\end{array}
$$

6. $250\% = 2.5 = \dfrac{250}{100} = 2\dfrac{50}{100} = 2\dfrac{1}{2}$

7. $34\% = 0.34 = \dfrac{34}{100} = \dfrac{17}{50}$

8. $7\% = 0.07 = \dfrac{7}{100}$

9. $75\% = 0.75 = \dfrac{75}{100} = \dfrac{3}{4}$

10. $400 \div 10 = 40$
    $40 \times 1 = 40$

11. $192 \div 6 = 32$
    $32 \times 5 = 160$

12. $96 \div 4 = 24$
    $24 \times 3 = 72$

13. $100 \div 5 = 20$
    $20 \times 4 = 80$

14. $6\dfrac{1}{3} \div 2\dfrac{1}{8} = \dfrac{19}{3} \div \dfrac{17}{8} =$
    $\dfrac{19}{3} \times \dfrac{8}{17} = \dfrac{152}{51} = 2\dfrac{50}{51}$

15. $8\dfrac{8}{10} \div 2\dfrac{2}{3} = \dfrac{88}{10} \div \dfrac{8}{3} =$
    $\dfrac{\overset{11}{88}}{10} \times \dfrac{3}{\underset{}{8}} = \dfrac{33}{10} = 3\dfrac{3}{10}$

16. 61 km
17. 110 mg

18. $\$79.97 \times 0.37 = \$29.5889$ off
    (rounded to $29.59)
    $\$79.97 - \$29.59 = \$50.38$ sale price

19. $\$8.95 + \$8.95 + \$12.49 = \$30.39$
    $\$30.39 \times 1.16 = \$35.2524$ total
    (rounded to $35.25)

20. $2\dfrac{2}{5}$ mi $\times 1\dfrac{2}{3}$ mi $= \dfrac{\overset{4}{12}}{5} \times \dfrac{5}{3} = 4$ mi

Systematic Review 15F

1. 0.005 g
2. 170,000 cm
3. 0.5 hl
4. Estimate:

$$
\begin{array}{r}
3 \\
\times\ .006 \\
\hline
.018
\end{array}
$$

$$
\begin{array}{r}
2952 \quad \text{3 places} \\
\times \qquad 6 \quad \text{3 places} \\
\hline
.017712 \quad \text{6 places}
\end{array}
$$

5. Estimate:

$$
\begin{array}{r}
100 \\
\times\ \ .6 \\
\hline
60
\end{array}
$$

$$
\begin{array}{r}
135123 \quad \text{3 places} \\
\times \qquad 612 \quad \text{3 places} \\
\hline
270246 \\
1\ 35123 \\
81\ 0738 \\
\hline
82.695276 \quad \text{6 places}
\end{array}
$$

6. $2\% = 0.02 = \dfrac{2}{100} = \dfrac{1}{50}$

7. $20\% = 0.2 = \dfrac{20}{100} = \dfrac{1}{5}$

8. $60\% = 0.6 = \dfrac{60}{100} = \dfrac{3}{5}$

9. $900\% = 9 = \dfrac{900}{100} = 9$

10. $54 \div 6 = 9$
    $9 \times 1 = 9$

11. $88 \div 8 = 11$
    $11 \times 3 = 33$

12. $490 \div 7 = 70$
    $70 \times 5 = 350$

13. $300 \div 5 = 60$
    $60 \times 2 = 120$

14. $5\frac{1}{9} \div 2\frac{1}{6} = \frac{46}{9} \div \frac{13}{6} =$

$\frac{46}{9_3} \times \frac{6^2}{13} = \frac{92}{39} = 2\frac{14}{39}$

15. $7\frac{1}{3} \div \frac{2}{5} = \frac{22}{3} \div \frac{2}{5} =$

$\frac{22^{11}}{3} \times \frac{5}{2} = \frac{55}{3} = 18\frac{1}{3}$

16. $8\frac{3}{4}$ lb $\div 5 = \frac{35}{4}$ lb $\div \frac{5}{1} =$

$\frac{35^7}{4}$ lb $\times \frac{1}{5} = \frac{7}{4}$ lb $= 1\frac{3}{4}$ lb

17. 0.5 kl

18. 68×2.54 cm $= 172.72$ cm $= 1,727.2$ mm

19. $\$48.00 \times 1.25 = \$60.00$ from Dad
$\$48.00 + \$60.00 = \$108.00$ total

20. $0.2 \times 100 = 20$ days

$\frac{1}{5} \times \frac{100}{1} = \frac{100}{5} = 20$ days

Lesson Practice 16A

1. done
2. done
3. $A \approx 3.14(1.6 \text{ m})^2 = 8.0384 \text{ m}^2$
$C \approx 2(3.14)(1.6 \text{ m}) = 10.048 \text{ m}$
4. $A \approx 3.14(15 \text{ ft})^2 = 706.5 \text{ ft}^2$
$C \approx (3.14)(30 \text{ ft}) = 94.2 \text{ ft}$
5. $A \approx 3.14(12 \text{ ft})^2 = 452.16 \text{ ft}^2$
6. $C \approx 2(3.14)(7.5 \text{ ft}) = 47.1 \text{ ft}$
7. $C \approx (3.14)(12 \text{ in}) = 37.68 \text{ in}$
37.68" > 36"; it will not be enough.
8. $A \approx 3.14(9 \text{ m})^2 = 254.34 \text{ m}^2$

Lesson Practice 16B

1. $A \approx 3.14(3 \text{ in})^2 = 28.26 \text{ in}^2$
$C \approx 2(3.14)(3 \text{ in}) = 18.84 \text{ in}$
2. $A \approx 3.14(5 \text{ ft})^2 = 78.5 \text{ ft}^2$
$C \approx (3.14)(10 \text{ ft}) = 31.4 \text{ ft}$
3. $A \approx 3.14(4.2 \text{ m})^2 = 55.3896 \text{ m}^2$
$C \approx 2(3.14)(4.2 \text{ m}) = 26.376 \text{ m}$

4. $A \approx 3.14(6 \text{ ft})^2 = 113.04 \text{ ft}^2$
$C \approx (3.14)(12 \text{ ft}) = 37.68 \text{ ft}$
5. $A \approx 3.14(16 \text{ ft})^2 = 803.84 \text{ ft}^2$
6. $C \approx 2(3.14)(11 \text{ ft}) = 69.08 \text{ ft}$
7. $C \approx (3.14)(5.4 \text{ mm}) = 16.956 \text{ mm}$
8. $A \approx 3.14(8.5 \text{ ft})^2 = 226.865 \text{ ft}^2$

Lesson Practice 16C

1. $A \approx 3.14(6 \text{ in})^2 = 113.04 \text{ in}^2$
$C \approx 2(3.14)(6 \text{ in}) = 37.68 \text{ in}$
2. $A \approx 3.14(7 \text{ ft})^2 = 153.86 \text{ ft}^2$
$C \approx (3.14)(14 \text{ ft}) = 43.96 \text{ ft}$
3. $A \approx 3.14(10 \text{ m})^2 = 314 \text{ m}^2$
$C \approx 2(3.14)(10 \text{ m}) = 62.8 \text{ m}$
4. $A \approx 3.14(4.75 \text{ ft})^2 = 70.84625 \text{ ft}^2$
$C \approx (3.14)(9.5 \text{ ft}) = 29.83 \text{ ft}$
5. $A \approx 3.14(6 \text{ in})^2 = 113.04 \text{ in}^2$
(rounded to 113 in^2)
6. $A \approx 3.14(9 \text{ ft})^2 = 254.34 \text{ ft}^2$
254.34 ft > 200 ft; it will not be enough.
7. $C \approx 2(3.14)(9 \text{ ft}) = 56.52 \text{ ft}$
56.52 ft $\div 2 = 28.26$ pieces
29 pieces must be bought
8. $C \approx 2(3.14)(100 \text{ mi}) = 628 \text{ miles}$

Systematic Review 16D

1. $A \approx 3.14(1.4 \text{ in})^2 = 6.1544 \text{ in}^2$
$C \approx 2(3.14)(1.4 \text{ in}) = 8.792 \text{ in}$
2. $A \approx 3.14(10 \text{ ft})^2 = 314 \text{ ft}^2$
$C \approx 3.14(20 \text{ ft}) = 62.8 \text{ ft}$
3. 0.15 m
4. 0.8 kl
5. 3,500 cg
6. $1.65 \times 9.43 = 15.5595$
7. $0.209 \times 8.07 = 1.68663$
8. $5.061 \times 3.94 = 19.94034$

9. $6\frac{1}{5} \div 1\frac{3}{4} = \frac{31}{5} \div \frac{7}{4} = \frac{31}{5} \times \frac{4}{7} =$

   $\frac{124}{35} = 3\frac{19}{35}$

10. $4\frac{3}{5} \div 4\frac{1}{2} = \frac{23}{5} \div \frac{9}{2} = \frac{23}{5} \times \frac{2}{9} =$

    $\frac{46}{45} = 1\frac{1}{45}$

11. $\frac{1}{2} \div \frac{1}{8} = \frac{1}{2} \times \frac{\cancel{8}^4}{1} = 4$

12. done

13. $10' + 10' + 10' + 10' = 40'$

14. $7\text{ m} + 15\text{ m} + 7\text{ m} + 15\text{ m} = 44\text{ m}$

15. $12\text{ m} + 13\text{ m} + 12\text{ m} + 13\text{ m} = 50\text{ m}$

16. $13\text{ m} + 12\text{ m} + 13\text{ m} = 38\text{ m}$

17. $38 \times 3\text{ ft} = 114\text{ ft}$ (approximately)

18. $\$18,550.00 \times 0.02 = \$371.00$

Systematic Review 16E

1. $A = 3.14(3.2\text{ cm})^2 = 32.1536\text{ cm}^2$
   $C = 2(3.14)(3.2\text{ cm}) = 20.096\text{ cm}$

2. $A = 3.14(2\text{ m})^2 = 12.56\text{ m}^2$
   $C = 3.14(4\text{ m}) = 12.56\text{ m}$

3. 0.01 km

4. 5 cl

5. $20,000\text{ dg}$

6. $0.2 \times 0.32 = 0.064$

7. $1.45 \times 0.04 = 0.058$

8. $0.005 \times 0.02 = 0.0001$

9. $3\frac{1}{2} \div 2\frac{1}{6} = \frac{7}{2} \div \frac{13}{6} =$

   $\frac{7}{\cancel{2}} \times \frac{\cancel{6}^3}{13} = \frac{21}{13} = 1\frac{8}{13}$

10. $5\frac{1}{4} \div 3\frac{1}{4} = \frac{21}{4} \div \frac{13}{4} =$

    $\frac{21}{\cancel{4}} \times \frac{\cancel{4}}{13} = \frac{21}{13} = 1\frac{8}{13}$

11. $\frac{7}{8} \div \frac{1}{16} = \frac{7}{\cancel{8}} \times \frac{\cancel{16}^2}{1} = 14$

12. $4" + 10" + 4" + 10" = 28"$

13. $45' + 45' + 45' + 45' = 180'$

14. $2.3\text{ m} + 1.2\text{ m} + 2.3\text{ m} + 1.2\text{ m} = 7\text{ m}$

15. $\$2,000.00 \times 0.23 = \$460.00$

16. $3.14(7\text{ in})^2 = 153.86\text{ in}^2$
    $\left(\text{round to } 154\text{ in}^2\right)$

17. $50\text{ m} + 75\text{ m} + 50\text{ m} + 75\text{ m} = 250\text{ m}$

18. $\$12.35/\text{m} \times 250\text{ m} = \$3,087.50$

19. 4 kl

20. $3\frac{1}{2} \div \frac{1}{4} = \frac{7}{2} \div \frac{1}{4} =$

    $\frac{7}{2} \times \frac{\cancel{4}^2}{1} = 14$ containers

Systematic Review 16F

1. $A \approx 3.14(1\text{ m})^2 = 3.14\text{ m}^2$
   $C \approx 2(3.14)(1\text{ m}) = 6.28\text{ m}$

2. $A \approx 3.14(0.5\text{ km})^2 = .785\text{ km}^2$
   $C \approx 3.14(1\text{ km}) = 3.14\text{ km}$

3. 1.7 cm

4. 0.32 L

5. $5,000\text{ g}$

6. $0.06 \times 0.05 = 0.003$

7. $45.1 \times 0.8 = 36.08$

8. $116 \times 0.19 = 22.04$

9. $7\frac{1}{8} \div 3\frac{3}{4} = \frac{57}{8} \div \frac{15}{4} =$

   $\frac{\cancel{57}^{19}}{\cancel{8}_2} \times \frac{\cancel{4}}{\cancel{15}_5} = \frac{19}{10} = 1\frac{9}{10}$

10. $2\frac{5}{6} \div 1\frac{1}{6} = \frac{17}{6} \div \frac{7}{6} =$

    $\frac{17}{\cancel{6}} \times \frac{\cancel{6}}{7} = \frac{17}{7} = 2\frac{3}{7}$

11. $\frac{5}{8} \div \frac{1}{4} = \frac{5}{\cancel{8}_2} \times \frac{\cancel{4}}{1} = \frac{5}{2} = 2\frac{1}{2}$

12. $6\text{ m} + 2\text{ m} + 6\text{ m} + 2\text{ m} = 16\text{ m}$

13. $1.2' + 1.2' + 1.2' + 1.2' = 4.8'$

14. $17" + 10" + 17" + 10" = 54"$

15. $\$10.95 \times 1.20 = \$13.14$

16. $3.14(15\text{ in})^2 = 706.5\text{ in}^2$

17. $\$20.50 + \$15.98 + \$11.24 = \$47.72$

18. $2\frac{1}{2}$ cups $\times 1\frac{1}{2} = \frac{5}{2}$ cups $\times \frac{3}{2}$

   $= \frac{15}{4}$ cups $= 3\frac{3}{4}$ cups

19. $8\frac{1}{4} \div \frac{3}{4} = \frac{\overset{11}{\cancel{33}}}{\cancel{4}} \times \frac{\cancel{4}}{\cancel{3}} = 11$ gifts

20. 100 m $= 100,000$ mm $= 0.1$ km

Lesson Practice 17A

1. done

2. done

3.
```
    .587
6 ) 3.522
    3 0
    ─────
     52
     48
    ─────
     42
     42
    ─────
```

4. $6 \times 0.587 = 3.522$

5.
```
    8.8
9 ) 79.2
    72
   ─────
    72
    72
   ─────
```

6. $9 \times 8.8 = 79.2$

7.
```
    .8
5 ) 4.0
    4 0
   ─────
```

8. $5 \times 0.8 = 4$

9.
```
    .004
2 ) .008
     8
```

10. $2 \times 0.004 = 0.008$

11.
```
    .0125
4 ) .0500
    4
   ─────
    10
     8
   ─────
    20
```

12. $4 \times 0.0125 = 0.05$

13. $3.6 \div 4 = 0.9$ pies

14. 5 pints $\div 8 = 0.625$ pints

15. $\$0.75 \div 5 = \$0.15$

Lesson Practice 17B

1.
```
    1.5
4 ) 6.0
    4
   ─────
    2 0
    2 0
   ─────
```

2. $4 \times 1.5 = 6$

3.
```
    .09
3 ) .27
    27
```

4. $3 \times 0.09 = 0.27$

5.
```
    .006
7 ) .042
    42
```

6. $7 \times 0.006 = 0.042$

7.
```
    .05
8 ) .40
    .40
```

8. $8 \times 0.05 = 0.4$

9.
```
    .203
6 ) 1.218
    1 2
   ─────
    018
     18
   ─────
```

10. $6 \times 0.203 = 1.218$

11.
```
     42.8
2 ) 85.6
    8
   ─────
    5
    4
   ─────
    1 6
    1 6
   ─────
```

12. $2 \times 42.8 = 85.6$

13. $24.8 \div 2 = 12.4$ pieces
    12 pieces with 0.4 of a
    piece left over

14. 2.4 bushels $\div 3 = 0.8$ bushels

15. 34.4 gal $\times 4$ qt/gal $= 137.6$ qt; 137 jars

Lesson Practice 17C

1.
$$\begin{array}{r} .08 \\ 5\overline{\smash)\,.40} \\ \underline{40} \end{array}$$

2. $5 \times 0.08 = 0.4$

3.
$$\begin{array}{r} .068 \\ 6\overline{\smash)\,.408} \\ \underline{36} \\ 48 \\ \underline{48} \end{array}$$

4. $6 \times 0.068 = 0.408$

5.
$$\begin{array}{r} .008 \\ 8\overline{\smash)\,.064} \\ \underline{64} \end{array}$$

6. $8 \times 0.008 = 0.064$

7.
$$\begin{array}{r} .3 \\ 4\overline{\smash)\,1.2} \\ \underline{12} \end{array}$$

8. $4 \times 0.3 = 1.2$

9.
$$\begin{array}{r} .5 \\ 6\overline{\smash)\,3.0} \\ \underline{30} \end{array}$$

10. $6 \times 0.5 = 3$

11.
$$\begin{array}{r} 3.12 \\ 9\overline{\smash)\,28.08} \\ \underline{27} \\ 10 \\ \underline{9} \\ 18 \\ \underline{18} \end{array}$$

12. $9 \times 3.12 = 28.08$

13. $345.5 \text{ miles} \div 2 \text{ days} = \dfrac{345.5 \text{ mi}}{2 \text{ days}}$

172.75 miles per day (mi/day)

14. $125.1 \text{ bushels} \div 9 \text{ days} = 13.9 \text{ bushels/day}$

15. $\$24.90 \div 3 = \$8.30 \text{ per child}$

Systematic Review 17D

1.
$$\begin{array}{r} .02 \\ 7\overline{\smash)\,.14} \\ \underline{.14} \end{array}$$

2. $7 \times 0.02 = 0.14$

3.
$$\begin{array}{r} 1.2 \\ 3\overline{\smash)\,3.6} \\ \underline{3} \\ 6 \\ \underline{6} \end{array}$$

4. $3 \times 1.2 = 3.6$

5.
$$\begin{array}{r} .061 \\ 9\overline{\smash)\,.549} \\ \underline{54} \\ 9 \\ \underline{9} \end{array}$$

6. $9 \times 0.061 = 0.549$

7.
$$\begin{array}{r} .625 \\ 8\overline{\smash)\,5.000} \\ \underline{4\ 8} \\ 20 \\ \underline{16} \\ 40 \\ \underline{40} \end{array}$$

8. $8 \times 0.625 = 5$

9. $3.14(2.5 \text{ mi})^2 = 19.625 \text{ mi}^2$

10. $3.14(5 \text{ mi}) = 15.7 \text{ mi}$

11. 0.015 hg

12. 0.005 kl

13. 160 cm

14. done

15. $13' + 3' + 13' + 3' = 32'$

16. $7.2 \text{ m} + 4.1 \text{ m} + 7.2 \text{ m} + 4.1 \text{ m} = 22.6 \text{ m}$

17. $\$32.50 \div 2 = \$16.25$

18. $2\dfrac{3}{4} - \dfrac{5}{8} = 2\dfrac{6}{8} - \dfrac{5}{8} = 2\dfrac{1}{8} \text{ pizza}$

Systematic Review 17E

1.
$$\begin{array}{r} .008 \\ 4\overline{\smash)\,.032} \\ \underline{32} \end{array}$$

2. $4 \times 0.008 = 0.032$

3.
$$\begin{array}{r} .05 \\ 6\overline{\smash)\,.30} \\ \underline{30} \end{array}$$

4. $6 \times 0.05 = 0.3$

5.
$$\begin{array}{r} 11.1 \\ 5\overline{)55.5} \\ \underline{5} \\ 5 \\ \underline{5} \\ 5 \end{array}$$

6. $5 \times 11.1 = 55.5$

7.
$$\begin{array}{r} .515 \\ 2\overline{)1.030} \\ \underline{1\ 0} \\ 3 \\ \underline{2} \\ 10 \\ \underline{10} \end{array}$$

8. $2 \times 0.515 = 1.030$

9. $3.14(8 \text{ km})^2 = 200.96 \text{ km}^2$

10. $2(3.14)(8 \text{ km}) = 50.24 \text{ km}$

11. 1 g

12. 0.003 km

13. 2.5 dl

14. $4\text{m} + 3.8\text{m} + 4\text{m} + 3.8\text{m} = 15.6 \text{ m}$

15. $25" + 9.1" + 25" + 9.1" = 68.2"$

16. $0.12 \text{ km} + 0.05 \text{ km} +$
$0.12 \text{ km} + 0.05 \text{ km} = 0.34 \text{ km}$

17. $\$15.00 \times 0.9 = \$13.50$ sale price

18. $\$13.50 \times 1.04 = \$14.04$ total cost
$\$15.00 - \$14.04 = \$0.96$ left over

19. $5.7 \text{ km} \div 3 = 1.9 \text{ km}$

20. $10 \text{ lb} - 2\frac{5}{8} \text{ lb} = 9\frac{8}{8} \text{ lb} - 2\frac{5}{8} \text{ lb} = 7\frac{3}{8} \text{ lb}$

Systematic Review 17F

1.
$$\begin{array}{r} .001 \\ 1\overline{).001} \\ \underline{1} \end{array}$$

2. $1 \times 0.001 = 0.001$

3.
$$\begin{array}{r} .32 \\ 7\overline{)2.24} \\ \underline{2\ 1} \\ 14 \\ \underline{14} \end{array}$$

4. $7 \times 0.32 = 2.24$

5.
$$\begin{array}{r} .05 \\ 3\overline{).15} \\ \underline{15} \end{array}$$

6. $3 \times 0.05 = 0.15$

7.
$$\begin{array}{r} .125 \\ 8\overline{)1.000} \\ \underline{8} \\ 20 \\ \underline{16} \\ 40 \\ \underline{40} \end{array}$$

8. $8 \times 0.125 = 1$

9. $3.14(2 \text{ mi})^2 = 12.56 \text{ mi}^2$

10. $2(3.14)(2 \text{ mi}) = 12.56 \text{ mi}$

11. 0.95 hl

12. 2 cm

13. 40 cg

14. $80" + 70" + 80" + 70" = 300"$

15. $3.1\text{m} + 1.3\text{m} + 3.1\text{m} + 1.3\text{m} = 8.8 \text{ m}$

16. $5\frac{1}{2}" + 3\frac{1}{2}" + 5\frac{1}{2}" + 3\frac{1}{2}" = 18"$

17. $5,000 \text{ m} = 5 \text{ km} = 500,000 \text{ cm}$

18. $\$5.76 - \$0.89 = \$4.87$

19. $\$38.40 \times 1.15 = \$44.16$ total
$\$44.16 \div 4 = \$11.04$ each

20. $5\frac{1}{2} - 2\frac{2}{3} = 5\frac{3}{6} - 2\frac{4}{6} =$
$4\frac{9}{6} - 2\frac{4}{6} = 2\frac{5}{6}$ rows

Lesson Practice 18A

1. done

2. done

3.
$$\begin{array}{r} 300 \\ 6\overline{)1800} \\ \underline{1800} \end{array}$$

4. $0.06 \times 300 = 18$

5.
$$\begin{array}{r} 1,000 \\ 5\overline{)5,000} \\ \underline{5,000} \end{array}$$

6. $0.5 \times 1,000 = 500$

7.
$$\begin{array}{r} 9,000 \\ 7\overline{)63,000} \\ \underline{63,000} \end{array}$$

8. $0.007 \times 9,000 = 63$

9.
$$\begin{array}{r} 120 \\ 9\overline{)1,080} \\ \underline{9} \\ 180 \\ \underline{180} \end{array}$$

10. $0.9 \times 120 = 108$

11.
$$\begin{array}{r} 1,600 \\ 25\overline{)40,000} \\ \underline{25} \\ 15,000 \\ \underline{15,000} \end{array}$$

12. $0.25 \times 1,600 = 400$

13.
$$\begin{array}{r} 80 \\ 5\overline{)400} \\ \underline{400} \end{array}$$

40 people ÷ 0.5 lb/person = 80 people

14.
$$\begin{array}{r} 80 \\ 3\overline{)240} \\ \underline{240} \end{array}$$

24 rows ÷ 0.3 rows/day = 80 days

15.
$$\begin{array}{r} 30 \\ 8\overline{)240} \\ \underline{240} \end{array}$$

24 rows ÷ 0.8 rows/day = 30 days

Lesson Practice 18B

1.
$$\begin{array}{r} 6,320 \\ 1\overline{)6,320} \\ 6,320 \end{array}$$

2. $0.1 \times 6,320 = 632$

3.
$$\begin{array}{r} 2,200 \\ 15\overline{)33,000} \\ \underline{30} \\ 3,000 \\ \underline{3,000} \end{array}$$

4. $0.15 \times 2,200 = 330$

5.
$$\begin{array}{r} 230 \\ 6\overline{)1,380} \\ \underline{1\ 2} \\ 180 \\ \underline{180} \end{array}$$

6. $0.6 \times 230 = 138$

7.
$$\begin{array}{r} 2,000 \\ 33\overline{)66,000} \\ \underline{66,000} \end{array}$$

8. $0.033 \times 2,000 = 66$

9.
$$\begin{array}{r} 530 \\ 4\overline{)2,120} \\ \underline{2,000} \\ 120 \\ \underline{120} \end{array}$$

(Filling in extra zeros to mark place value is optional when working a division problem.)

10. $0.4 \times 530 = 212$

11.
$$\begin{array}{r} 500 \\ 75\overline{)37,500} \\ \underline{37,500} \end{array}$$

12. $0.75 \times 500 = 375$

13. $\$200 \div \$0.50 = 400$ gifts

14. 36 yd ÷ 0.9 yd/pillow = 40 pillows

15. 3 tons ÷ 0.01 tons/day = 300 days

Lesson Practice 18C

1.
$$\begin{array}{r} 2,290 \\ 2\overline{)4,580} \\ \underline{4,000} \\ 580 \\ \underline{400} \\ 180 \\ \underline{180} \end{array}$$

2. $0.2 \times 2,290 = 458$

3.
$$\begin{array}{r} 200 \\ 8\overline{)1,600} \\ \underline{1,600} \end{array}$$

4. $0.08 \times 200 = 16$

5.
$$\begin{array}{r} 160 \\ 7\overline{)1,120} \\ \underline{700} \\ 420 \\ \underline{420} \end{array}$$

6. $0.7 \times 160 = 112$

7.
$$\begin{array}{r} 38,000 \\ 9\overline{)342,000} \\ \underline{270,000} \\ 72,000 \\ \underline{72,000} \end{array}$$

8. $0.009 \times 38,000 = 342$

9.
$$\begin{array}{r} 1,190 \\ 3\overline{)3,570} \\ \underline{3,000} \\ 570 \\ \underline{300} \\ 270 \\ \underline{270} \end{array}$$

10. $0.3 \times 1,190 = 357$

11.
$$\begin{array}{r} 1,000 \\ 45\overline{)45,000} \\ \underline{45,000} \end{array}$$

12. $0.45 \times 1,000 = 450$

13. 5 L $\div$ 0.05 L/batch = 100 batches

14. \$35.00 $\div$ \$0.25 / quarter = 140 quarters

15. 224 lb $\div$ 4 lb/cust. = 560 customers

Systematic Review 18D

1.
$$\begin{array}{r} 60 \\ 6\overline{)360} \\ \underline{360} \end{array}$$

2. $0.6 \times 60 = 36$

3.
$$\begin{array}{r} 2,000 \\ 9\overline{)18,000} \\ \underline{18,000} \end{array}$$

4. $0.09 \times 2,000 = 180$

5.
$$\begin{array}{r} 3.54 \\ 2\overline{)7.08} \\ \underline{6\ 00} \\ 10 \\ \underline{10} \\ 8 \\ \underline{8} \end{array}$$

6. $2 \times 3.54 = 7.08$

7.
$$\begin{array}{r} .002 \\ 15\overline{).030} \\ \underline{30} \end{array}$$

8. $15 \times 0.002 = .03$

9. $3.14(5.5 \text{ ft})^2 = 94.985 \text{ ft}^2$

10. $3.14(11 \text{ ft}) = 34.54 \text{ ft}$

11. $8 \times 8 = 64$

12. $10 \times 10 = 100$

13. $2 \times 2 = 4$

14. done

15. $3' + 4' + 5' = 12'$

16. $5.4' + 3.9' + 7.8' = 17.1'$

17. $14" + 14" + 14" + 14" = 56"$

18. 5 tons $\times 1.50 = 7.5$ tons

Systematic Review 18E

1.
$$\begin{array}{r} 1,000 \\ 1\overline{)1,000} \\ \underline{1,000} \end{array}$$

2. $0.1 \times 1,000 = 100$

3.
$$\begin{array}{r} 250 \\ 6\overline{)1,500} \\ \underline{1\ 2} \\ 300 \\ \underline{300} \end{array}$$

4. $0.06 \times 250 = 15$

5.
$$\begin{array}{r} 1.9 \\ 8\overline{)15.2} \\ \underline{8\ 0} \\ 7\ 2 \\ \underline{7\ 2} \end{array}$$

6. $8 \times 1.9 = 15.2$

7.

$$30 \overline{\smash{\big)}\,.0060}$$
$$.0002$$
$$\underline{60}$$

8. $30 \times 0.0002 = 0.006$

9. $3.14(9 \text{ m})^2 = 254.34 \text{ m}^2$

10. $2(3.14)(9 \text{ m}) = 56.52 \text{ m}$

11. $12 \times 12 = 144$

12. $1 \times 1 = 1$

13. $5 \times 5 = 25$

14. $8' + 8' + 10' = 26'$

15. $5" + 12" + 13" = 30"$

16. $1.7 \text{ m} + 1.1 \text{ m} + 1.9 \text{ m} = 4.7 \text{ m}$

17. $20 \text{ cm} + 30 \text{ cm} + 20 \text{ cm} + 30 \text{ cm} = 100 \text{ cm} = 1 \text{ m}$

18. $100 \times 2.1 = 210$ nature books

19. $\frac{1}{2}c + 1\frac{1}{4}c = \frac{2}{4}c + 1\frac{1}{4}c = 1\frac{3}{4}$ cups

20. $\$8.00 \div \$0.75 = 10.67 = 10$ cones

Systematic Review 18F

1.

$$3 \overline{\smash{\big)}\,369,000}$$
$$123,000$$
$$\underline{300,000}$$
$$69,000$$
$$\underline{60,000}$$
$$9,000$$
$$\underline{9,000}$$

2. $0.003 \times 123,000 = 369$

3.

$$49 \overline{\smash{\big)}\,4,900}$$
$$100$$
$$\underline{4,900}$$

4. $0.49 \times 100 = 49$

5.

$$5 \overline{\smash{\big)}\,22.5}$$
$$4.5$$
$$\underline{200}$$
$$25$$
$$\underline{25}$$

6. $5 \times 4.5 = 22.5$

7.

$$8 \overline{\smash{\big)}\,.0016}$$
$$.0002$$
$$\underline{16}$$

8. $8 \times 0.0002 = 0.0016$

9. $3.14(1.2 \text{ in})^2 = 4.5216 \text{ in}^2$

10. $2(3.14)(1.2 \text{ in}) = 7.536 \text{ in}$

11. $25 \times 25 = 625$

12. $7 \times 7 = 49$

13. $100 \times 100 = 10,000$

14. $25' + 25' + 32' = 82'$

15. $6" + 8" + 10" = 24"$

16. $0.89 \text{ m} + 0.7 \text{ m} + 1.08 \text{ m} = 2.67 \text{ m}$

17. $25 \text{ lb} \div 0.5 \text{ lb/pack.} = 50$ packages

18. $\$37.50 \times 1.50 = \$56.25$
$\$56.25 \div 50 \text{ pack.} = \$1.125 / \text{package}$
rounds to $\$1.13$ per package

19. $\$42.50 \times 1.34 = \$56.95$

20. $4,520 \text{ g} = 4.52 \text{ kg}$

Lesson Practice 19A

1. done

2. done

3. $0.06X = 24$
$$\frac{0.06X}{0.06} = \frac{24}{0.06}$$
$$X = 24 \div 0.06 = 400$$

4. $0.06(400) = 24$
$$24 = 24$$

5. $0.7A = 490$
$$\frac{0.7A}{0.7} = \frac{490}{0.7}$$
$$A = 490 \div 0.7 = 700$$

6. $0.7(700) = 490$
$$490 = 490$$

7. $0.002R = 6$
$$\frac{0.002R}{0.002} = \frac{6}{0.002}$$
$$R = 6 \div 0.002 = 3,000$$

8. $0.002(3,000) = 6$
$$6 = 6$$

9. $0.2M = \$84$
$$\frac{0.2M}{0.2} = \frac{\$84}{0.2}$$
$$M = \$84 \div 0.2 = \$420.00$$

10. $0.15T = 3'$

$$\frac{0.15T}{0.15} = \frac{3'}{0.15}$$

$$T = 3' \div 0.15 = 20'$$

Lesson Practice 19B

1. $0.5Y = 8$

$$\frac{0.5Y}{0.5} = \frac{8}{0.5}$$

$$Y = 8 \div 0.5 = 16$$

2. $0.5(16) = 8$

$$8 = 8$$

3. $0.03X = 21$

$$\frac{0.03X}{0.03} = \frac{21}{0.03}$$

$$X = 21 \div 0.03 = 700$$

4. $0.03(700) = 21$

$$21 = 21$$

5. $0.1A = 30$

$$\frac{0.1A}{0.1} = \frac{30}{0.1}$$

$$A = 30 \div 0.1 = 300$$

6. $0.1(300) = 30$

$$30 = 30$$

7. $0.008R = 16$

$$\frac{0.008R}{0.008} = \frac{16}{0.008}$$

$$R = 16 \div 0.008 = 2,000$$

8. $0.008(2,000) = 16$

$$16 = 16$$

9. $0.7P = \$140$

$$\frac{0.7P}{0.7} = \frac{\$140}{0.7}$$

$$P = \$140 \div 0.7 = \$200.00$$

10. $0.8H = 60''$

$$\frac{0.8H}{0.8} = \frac{60''}{0.8}$$

$$H = 60'' \div 0.8 = 75''$$

Lesson Practice 19C

1. $0.6Y = 6$

$$\frac{0.6Y}{0.6} = \frac{6}{0.6}$$

$$Y = 6 \div 0.6 = 10$$

2. $0.6(10) = 6$

$$6 = 6$$

3. $0.04X = 32$

$$\frac{0.04X}{0.04} = \frac{32}{0.04}$$

$$X = 32 \div 0.04 = 800$$

4. $0.04(800) = 32$

$$32 = 32$$

5. $0.3A = 93$

$$\frac{0.3A}{0.3} = \frac{93}{0.3}$$

$$A = 93 \div 0.3 = 310$$

6. $0.3(310) = 93$

$$93 = 93$$

7. $0.005R = 5$

$$\frac{0.005R}{0.005} = \frac{5}{0.005}$$

$$R = 5 \div 0.005 = 1,000$$

8. $0.005(1,000) = 5$

$$5 = 5$$

9. $0.25G = 4 \text{ bushels}$

$$\frac{0.25G}{0.25} = \frac{4 \text{ bushels}}{0.25}$$

$$G = 4 \text{ bushels} \div 0.25$$

$$G = 16 \text{ bushels}$$

10. $0.05R = \$10$

$$\frac{0.05R}{0.05R} = \frac{\$10}{0.05}$$

$$R = \$10 \div 0.05 = \$200.00$$

Systematic Review 19D

1. $0.22Q = 88$

$$\frac{0.22Q}{0.22} = \frac{88}{0.22}$$

$$Y = 88 \div 0.22 = 400$$

2. $0.22(400) = 88$
   $88 = 88$

3. $0.007X = 35$
   $$\frac{0.007X}{0.007} = \frac{35}{0.007}$$
   $X = 35 \div 0.007 = 5{,}000$

4. $0.007(5{,}000) = 35$
   $35 = 35$

5. $4 \div 20 = 0.2$

6. $20 \times 0.2 = 4$

7. $3.6 \div 18 = 0.2$

8. $18 \times 0.2 = 3.6$

9. done

10. $10 \text{ ft} \times 10 \text{ ft} = 100 \text{ ft}^2$

11. $15 \text{ m} \times 7 \text{ m} = 105 \text{ m}^2$

12. $9 \text{ ft} \times 12 \text{ ft} = 108 \text{ ft}^2$

13. $9' + 12' + 9' + 12' = 42'$

14. $0.35M = \$105$
    $$\frac{0.35M}{0.35} = \frac{\$105}{0.35}$$
    $M = \$105 \div 0.35 = \$300.00$

15. $2(3.14)(5 \text{ mi}) = 31.4 \text{ mi}$

16. $3.14(5 \text{ mi})^2 = 78.5 \text{ mi}^2$

17. $5 \text{ km} = 5{,}000 \text{ m}$
    $5{,}000 \text{ m} > 500 \text{ m}$; She drove further.

18. $\$30.75 \div 5 = \$6.15$

Systematic Review 19E

1. $0.9D = 27$
   $$\frac{0.9D}{0.9} = \frac{27}{0.9}$$
   $D = 27 \div 0.9 = 30$

2. $0.9(30) = 27$
   $27 = 27$

3. $0.08F = 5$
   $$\frac{0.08F}{0.08} = \frac{5}{0.08}$$
   $F = 5 \div 0.08 = 62.5$

4. $0.08(62.5) = 5$
   $5 = 5$

5. $5{,}500 \div 11 = 500$

6. $0.11 \times 500 = 55$

7. $0.16 \div 40 = 0.004$

8. $40 \times 0.004 = 0.16$

9. $10 \text{ in} \times 4 \text{ in} = 40 \text{ in}^2$

10. $45 \text{ ft} \times 45 \text{ ft} = 2{,}025 \text{ ft}^2$

11. $2.3 \text{ m} \times 1.2 \text{ m} = 2.76 \text{ m}^2$

12. $4 \text{ yd} \times 11 \text{ yd} = 44 \text{ yd}^2$

13. $3 \text{ ft} + 3 \text{ ft} + 3 \text{ ft} + 3 \text{ ft} = 12 \text{ ft}$
    $12 \times 1 = 12$ flowers

14. $0.40P = \$80$
    $$\frac{0.40P}{0.40} = \frac{\$80}{0.40}$$
    $P = \$80 \div 0.40 = \$200.00$

15. $10 \text{ in} \times 12 \text{ in} = 120 \text{ in}^2$

16. $3.14(6 \text{ in})^2 = 113.04 \text{ in}^2$

17. $120 \text{ in}^2 > 113.04 \text{ in}^2$
    The rectangular one would be better.

18. $2 \text{ m} = 200 \text{ cm}$
    They are the same height.

19. $\$57.04 \div 8 = \$7.13$

20. $4.5" + 5" + 3.75" = 13.25"$

Systematic Review 19F

1. $0.32G = 64$
   $$\frac{0.32G}{0.32} = \frac{64}{0.32}$$
   $G = 64 \div 0.32 = 200$

2. $0.32(200) = 64$
   $64 = 64$

3. $0.005Y = 4$
   $$\frac{0.005Y}{0.005} = \frac{4}{0.005}$$
   $Y = 4 \div 0.005 = 800$

4. $0.005(800) = 4$
   $4 = 4$

5. $10{,}000 \div 1 = 10{,}000$

6. $0.001 \times 10{,}000 = 10$

7. $1.26 \div 21 = 0.06$

8. $0.06 \times 21 = 1.26$

9. $6 \text{ m} \times 2 \text{ m} = 12 \text{ m}^2$
10. $1.2 \text{ ft} \times 1.2 \text{ ft} = 1.44 \text{ ft}^2$
11. $17 \text{ in} \times 10 \text{ in} = 170 \text{ in}^2$
12. $5 \text{ ft} \times 6 \text{ ft} = 30 \text{ ft}^2$
13. $3.14(3 \text{ ft})^2 = 28.26 \text{ ft}^2$
14. $30 \text{ ft}^2 - 28.26 \text{ ft}^2 = 1.74 \text{ ft}^2$
15. $1 \text{ mi} + 1 \text{ mi} + 1 \text{ mi} + 1 \text{ mi} = 4 \text{ mi}$
16. $0.32T = 200 \text{ hr}$

$$\frac{0.32T}{0.32} = \frac{200 \text{ hr}}{0.32}$$

$$T = 200 \text{ hr} \div 0.32 = 625 \text{ hr}$$

17. $625 \text{ hr} - 200 \text{ hr} = 425 \text{ hr}$
18. $652 \text{ g} = 0.652 \text{ kg}$
19. $\$121.92 \div 12 = \$10.16$
20. $6.25'' + 7.1'' + 6.25'' + 7.1'' = 26.7''$

Lesson Practice 20A

1. done
2. done
3. done
4. done
5.
```
      9
   9 |81
      81
```
6. $0.9 \times 9 = 8.1$
7.
```
      40
   6 |240
      240
```
8. $0.006 \times 40 = 0.24$
9.
```
      59
   8 |472
      472
```
10. $0.8 \times 59 = 47.2$
11.
```
       2.72
   15 |40.80
       30 00
       10 80
       10 50
          30
```

12. $1.5 \times 2.72 = 4.080$
13. $1.8 \text{ L} \div 0.3 \text{ L/guest} = 6 \text{ guests}$
14. $2.6 \text{ acres} \div 1.3 \text{ acres/day} = 2 \text{ days}$
15. $6.9 \text{ pt} \div 2.3 \text{ pt/person} = 3 \text{ people}$

Lesson Practice 20B

1.
```
       .09
   5 |.45
      .45
```
2. $0.5 \times 0.09 = 0.045$
3.
```
        5
   82 |410
       410
```
4. $0.82 \times 5 = 4.1$
5.
```
       1.1
   2 |2.2
      2 0
        2
        2
```
6. $0.2 \times 1.1 = 0.22$
7.
```
      71
   9 |639
      630
        9
        9
```
8. $0.009 \times 71 = 0.639$
9.
```
       61.2
   7 |428.4
      420 0
        8 4
        7 0
        1 4
        1 4
```
10. $0.7 \times 61.2 = 42.84$
11.
```
       15
   21 |315
       210
       105
       105
```
12. $2.1 \times 15 = 31.5$
13. $0.856 \text{ kg} \div 0.4 \text{ kg/sack} = 2.14 \text{ sacks}$
    He has 3 sacks.

14. 5.5 lb ÷ 0.25 lb/favor = 22 favors
15. $45.50 ÷ $0.35 per item = 130 items

Lesson Practice 20C

1.
$$
\begin{array}{r}
.06 \\
8\overline{)\,.48} \\
\underline{.48}
\end{array}
$$

2. $0.8 \times 0.06 = 0.048$

3.
$$
\begin{array}{r}
40.6 \\
22\overline{)\,893.2} \\
\underline{8800} \\
13\ 2 \\
\underline{13\ 2}
\end{array}
$$

4. $0.22 \times 40.6 = 8.932$

5.
$$
\begin{array}{r}
13 \\
5\overline{)\,65} \\
\underline{50} \\
15 \\
\underline{15}
\end{array}
$$

6. $0.5 \times 13 = 6.5$

7.
$$
\begin{array}{r}
580 \\
7\overline{)\,4060} \\
\underline{3500} \\
560 \\
\underline{560}
\end{array}
$$

8. $0.007 \times 580 = 4.06$

9.
$$
\begin{array}{r}
91.2 \\
4\overline{)\,364.8} \\
\underline{360\ 0} \\
4\ 8 \\
\underline{4\ 8}
\end{array}
$$

10. $0.4 \times 91.2 = 36.48$

11.
$$
\begin{array}{r}
16 \\
32\overline{)\,512} \\
\underline{320} \\
192 \\
\underline{192}
\end{array}
$$

12. $3.2 \times 16 = 51.2$
13. 25.5 lb ÷ 0.5 lb/bag = 51 bags
14. 4.6 mi ÷ 0.2 mi/rest = 23 rests
15. $16.80 ÷ $2.10 / item = 8 items

Systematic Review 20D

1.
$$
\begin{array}{r}
.5 \\
3\overline{)\,1.5} \\
\underline{1\ 5}
\end{array}
$$

2. $0.3 \times 0.5 = 0.15$

3.
$$
\begin{array}{r}
1.6 \\
25\overline{)\,40.0} \\
\underline{25\ 0} \\
15\ 0 \\
\underline{15\ 0}
\end{array}
$$

4. $0.025 \times 1.6 = 0.04$

5.
$$
\begin{array}{r}
.002 \\
60\overline{)\,.120} \\
\underline{120}
\end{array}
$$

6. $60 \times 0.002 = 0.12$

7.
$$
\begin{array}{r}
430 \\
1\overline{)\,430} \\
\underline{400} \\
30 \\
\underline{30}
\end{array}
$$

8. $0.1 \times 430 = 43$

9. $0.8G = 40$
   $G = 40 \div 0.8 = 50$

10. $0.15Y = 0.3$
    $Y = 0.3 \div 0.15 = 2$

11. $8\frac{1}{2} + 2\frac{7}{8} = 8\frac{4}{8} + 2\frac{7}{8} = 10\frac{11}{8} = 11\frac{3}{8}$

12. $16 - 1\frac{2}{3} = 15\frac{3}{3} - 1\frac{2}{3} = 14\frac{1}{3}$

13. $3\frac{1}{8} \times 1\frac{1}{3} = \frac{25}{{}_2\cancel{8}} \times \frac{\cancel{4}}{3} = \frac{25}{6} = 4\frac{1}{6}$

14. done

15. 13 in × 28 in = 364 in^2

16. 2 m × 1.2 m = 2.4 m^2

17. 220m + 150m + 220m + 150m = 740 m
    220m × 100m = 22,000 m^2

18. $0.25M = 37.50
    $M = $37.50 \div 0.25 = 150.00

Systematic Review 20E

1.
$$\begin{array}{r} 228 \\ 2\,\overline{)456} \\ \underline{400} \\ 56 \\ \underline{40} \\ 16 \\ \underline{16} \end{array}$$

2. $0.02 \times 228 = 4.56$

3.
$$\begin{array}{r} 910 \\ 9\,\overline{)8,190} \\ \underline{8,100} \\ 90 \\ \underline{90} \end{array}$$

4. $0.009 \times 910 = 8.19$

5.
$$\begin{array}{r} 5.01 \\ 12\,\overline{)60.12} \\ \underline{60\ 00} \\ 12 \\ \underline{12} \end{array}$$

6. $12 \times 5.01 = 60.12$

7.
$$\begin{array}{r} 20 \\ 45\,\overline{)900} \\ \underline{900} \end{array}$$

8. $4.5 \times 20 = 90$

9. $6.2X = 7.44$
$X = 7.44 \div 6.2 = 1.2$

10. $0.4B = 0.88$
$B = 0.88 \div 0.4 = 2.2$

11. $6\frac{3}{8} \div 3\frac{1}{4} =$

$\frac{51}{8} \div \frac{13}{4} =$

$\frac{51}{\underset{2}{\cancel{8}}} \times \frac{\cancel{4}}{13} = \frac{51}{26} = 1\frac{25}{26}$

12. $65 - 21\frac{5}{6} = 64\frac{6}{6} - 21\frac{5}{6} = 43\frac{1}{6}$

13. $9\frac{3}{10} \times 5\frac{2}{3} =$

$\frac{\overset{31}{\cancel{93}}}{10} \times \frac{17}{\cancel{3}} = \frac{527}{10} = 52\frac{7}{10}$

14. $3.8 \text{ m} \times 3 \text{ m} = 11.4 \text{ m}^2$

15. $20 \text{ mm} \times 7 \text{ mm} = 140 \text{ mm}^2$

16. $25 \text{ in} \times 8.5 \text{ in} = 212.5 \text{ in}^2$

17. $25 \text{ yd} \times 33 \text{ yd} = 825 \text{ yd}^2$ total
$825 \text{ yd}^2 \times \frac{1}{3} = 275 \text{ yd}^2$ planted

18. $275 \text{ yd}^2 \div 0.5 \text{ yd}^2 / \text{plant} = 550$ plants

19. $550 \text{ plants} \times \$0.85 / \text{plant} = \$467.50$

20. $1.05A = \$67.41$
$A = \$67.41 \div 1.05 = \$64.20$

Systematic Review 20F

1.
$$\begin{array}{r} 78 \\ 4\,\overline{)312} \\ \underline{280} \\ 32 \\ \underline{32} \end{array}$$

2. $0.4 \times 78 = 31.2$

3.
$$\begin{array}{r} 15 \\ 28\,\overline{)420} \\ \underline{280} \\ 140 \\ \underline{140} \end{array}$$

4. $0.28 \times 15 = 4.2$

5.
$$\begin{array}{r} .056 \\ 35\,\overline{)1.960} \\ \underline{1\ 750} \\ 210 \\ \underline{210} \end{array}$$

6. $35 \times 0.056 = 1.96$

7.
$$\begin{array}{r} 20 \\ 88\,\overline{)1760} \\ \underline{1760} \end{array}$$

8. $8.8 \times 20 = 176$

9. $0.11Q = 0.55$
$Q = 0.55 \div 0.11 = 5$

10. $1.3W = 0.39$
$W = 0.39 \div 1.3 = 0.3$

11. $\frac{6}{5} \div \frac{2}{5} = \frac{6 \div 2}{1} = 3$

12. $18 + 6\frac{7}{8} = 24\frac{7}{8}$

13. $\frac{11}{\underset{3}{\cancel{12}}} \times \frac{\cancel{4}}{5} = \frac{11}{15}$

14. $80 \text{ in} \times 55 \text{ in} = 4{,}400 \text{ in}^2$

15. $3.1 \text{ m} \times 0.9 \text{ m} = 2.79 \text{ m}^2$

16. $\dfrac{11}{2} \text{ in} \times \dfrac{21}{8} \text{ in} = \dfrac{231}{16} \text{ in}^2 = 14 \dfrac{7}{16} \text{ in}^2$

17. $\pi (6 \text{ ft})^2 \approx 113.04 \text{ ft}^2$

18. $5 \cancel{\text{ km}} \times \dfrac{1{,}000 \text{ m}}{1 \cancel{\text{ km}}} = 5{,}000 \text{ m}$

19. $1.15M = \$18.86$
$M = \$18.86 \div 1.15 = \$16.40$

20. $9 \text{ mi} + 12 \text{ mi} + 15 \text{ mi} = 36 \text{ mi}$

Lesson Practice 21A

1. done

2.
```
      .712
  5 |3.560
    3 500
      60
      50
      10
      10
```

3.
```
     11.25
  8 |90.00
   80 00
   10 00
    8 00
    2 00
    1 60
      40
      40
```

4. done

5.
```
    1.973 ≈ 1.97
  3 |5.920
   3 000
   2 920
   2 700
     220
     210
      10
```

(≈ means "approximately equal to")

6.
```
      17.555 ≈ 17.56
  9 |158.000
    90 000
    68 000
    63 000
     5 000
     4 500
       500
       450
        50
        45
         5
```

7. done

8.
```
      7.33  or 7.3̄
  3 |22.00
    21 00
     1 00
       90
       10
        9
```

9.
```
      9.166  or 9.16̄
  6 |55.000
    54 000
     1 000
       600
       400
       360
        40
        36
```

10. done

11.
```
      .59 6/8 = 0.59 3/4
  8 |4.78
    4 00
      78
      72
       6
```

12.
$$
\begin{array}{r}
4.72\frac{2}{9} \\
9\overline{)42.50} \\
\underline{36\ 00} \\
6\ 50 \\
\underline{6\ 30} \\
20 \\
\underline{18} \\
2
\end{array}
$$

Lesson Practice 21B

1.
$$
\begin{array}{r}
.2335 \\
2\overline{).4670} \\
\underline{4000} \\
670 \\
\underline{600} \\
70 \\
\underline{60} \\
10 \\
\underline{10}
\end{array}
$$

2.
$$
\begin{array}{r}
12.75 \\
4\overline{)51.00} \\
\underline{40\ 00} \\
11\ 00 \\
\underline{8\ 00} \\
3\ 00 \\
\underline{280} \\
20 \\
\underline{20}
\end{array}
$$

3.
$$
\begin{array}{r}
15.4 \\
5.\overline{)77.0} \\
\underline{5\ 00} \\
2\ 70 \\
\underline{2\ 50} \\
20 \\
\underline{20}
\end{array}
$$

4.
$$
\begin{array}{r}
4.557 \approx 4.56 \\
7\overline{)31.900} \\
\underline{28\ 000} \\
3\ 900 \\
\underline{3\ 500} \\
400 \\
\underline{350} \\
50 \\
\underline{49}
\end{array}
$$

5.
$$
\begin{array}{r}
1.093 \approx 1.09 \\
9\overline{)9.840} \\
\underline{9\ 000} \\
840 \\
\underline{810} \\
30 \\
\underline{27}
\end{array}
$$

6.
$$
\begin{array}{r}
50.833 \approx 50.83 \\
6\overline{)305.000} \\
\underline{300\ 000} \\
5\ 000 \\
\underline{4\ 800} \\
200 \\
\underline{180} \\
20 \\
\underline{18}
\end{array}
$$

7.
$$
\begin{array}{r}
.13\overline{3} \text{ or } 0.1\overline{3} \\
3\overline{).400} \\
\underline{300} \\
100 \\
\underline{90} \\
10 \\
\underline{9}
\end{array}
$$

8.
```
      .01428571 or 0.0142857
 7 ).10000000
        7000000
        3000000
        2800000
         200000
         140000
          60000
          56000
           4000
           3500
            500
            490
             10
              7
```

9.
```
      5.9166 or 5.916
 6 )35.5000
     30 0000
      5 5000
      5 4000
        1000
         600
         400
         360
          40
          36
```

10.
```
      14.07 1/7
 7 )98.50
     70 00
     28 50
     28 00
        50
        49
         1
```

11.
```
      244.66 2/3
 3 )734.00
     600 00
     134 00
     120 00
      14 00
      12 00
       2 00
       1 80
```

12.
```
      13.87 4/8 = 13.84 1/2     13.87 1/2
 8 )111.00
     80 00
     31 00
     24 00
      7 00
      6 40
        60
        56
         4
```

Lesson Practice 21C

1.
```
      7.1375
 8 )57.1000
     56 0000
      1 1000
        8000
        3000
        2400
         600
         560
          40
          40
```

2.
```
      19.5
 2 )39.0
     200
     190
     180
      10
      10
```

3.
```
      3.125
 4 )12.500
     12 000
        500
        400
        100
         80
         20
         20
```

4.
$$\begin{array}{r} 2.266 \approx 2.27 \\ 3\overline{)6.800} \\ \underline{6\,000} \\ 800 \\ \underline{600} \\ 200 \\ \underline{180} \\ 20 \\ \underline{18} \end{array}$$

5.
$$\begin{array}{r} .742 \approx 0.74 \\ 7\overline{)5.200} \\ \underline{4\,900} \\ 300 \\ \underline{280} \\ 20 \\ \underline{14} \end{array}$$

6.
$$\begin{array}{r} .535 \approx 0.54 \\ 9\overline{)4.820} \\ \underline{4\,500} \\ 320 \\ \underline{270} \\ 50 \\ \underline{45} \end{array}$$

7.
$$\begin{array}{r} .4077 \text{ or } 0.40\overline{7} \\ 9\overline{)3.6700} \\ \underline{3\,6000} \\ 700 \\ \underline{630} \\ 70 \\ \underline{63} \end{array}$$

8.
$$\begin{array}{r} .181 \text{ or } 0.1\overline{8} \\ 11\overline{)2.000} \\ \underline{1\,100} \\ 900 \\ \underline{880} \\ 20 \\ \underline{11} \end{array}$$

9.
$$\begin{array}{r} 15.0166 \text{ or } 15.01\overline{6} \\ 6\overline{)90.1000} \\ \underline{60\,0000} \\ 30\,1000 \\ \underline{30\,0000} \\ 1000 \\ \underline{600} \\ 400 \\ \underline{360} \\ 40 \\ \underline{36} \end{array}$$

10.
$$\begin{array}{r} .78\dfrac{2}{10} = 0.78\dfrac{1}{5} \\ 10\overline{)7.82} \\ \underline{7\,00} \\ 82 \\ \underline{80} \\ 2 \end{array}$$

11.
$$\begin{array}{r} .04\dfrac{5}{7} \\ 7\overline{).33} \\ \underline{28} \\ 5 \end{array}$$

12.
$$\begin{array}{r} 2.04\dfrac{2}{12} = 2.04\dfrac{1}{6} \\ 12\overline{)24.50} \\ \underline{24\,00} \\ 50 \\ \underline{48} \\ 2 \end{array}$$

Systematic Review 21D

1.
$$\begin{array}{r} .178 \approx 0.18 \\ 7\overline{)1.250} \\ \underline{700} \\ 550 \\ \underline{490} \\ 60 \\ \underline{56} \end{array}$$

2.
$$
\begin{array}{r}
1.666 \approx 1.67 \\
3\overline{)5.000} \\
\underline{3\ 000} \\
2\ 000 \\
\underline{1\ 800} \\
200 \\
\underline{180} \\
20 \\
\underline{18}
\end{array}
$$

3.
$$
\begin{array}{r}
5.483 \approx 5.48 \\
6\overline{)32.900} \\
\underline{30\ 000} \\
2\ 900 \\
\underline{2\ 400} \\
500 \\
\underline{480} \\
20 \\
\underline{18}
\end{array}
$$

4.
$$
\begin{array}{r}
.9344 \text{ or } 0.93\overline{4} \\
9\overline{)8.4100} \\
\underline{8\ 1000} \\
3100 \\
\underline{2700} \\
400 \\
\underline{360} \\
40 \\
\underline{36}
\end{array}
$$

5.
$$
\begin{array}{r}
6.33 \text{ or } 6.\overline{3} \\
6\overline{)38.00} \\
\underline{36\ 00} \\
2\ 00 \\
\underline{1\ 80} \\
20 \\
\underline{18}
\end{array}
$$

6.
$$
\begin{array}{r}
.2857142 \text{ or } 0.\overline{285714} \\
7\overline{)2.0000000} \\
\underline{1\ 4000000} \\
6000000 \\
\underline{5600000} \\
400000 \\
\underline{350000} \\
50000 \\
\underline{49000} \\
1000 \\
\underline{700} \\
300 \\
\underline{280} \\
20 \\
\underline{14}
\end{array}
$$

7. $8X = 0.5$

$X = 0.5 \div 8 = 0.06\dfrac{2}{8} = 0.06\dfrac{1}{4}$

8. $0.3Y = 5.63$

$Y = 5.63 \div 0.3 = 18.76\dfrac{2}{3}$

9. $10F = 0.46$

$F = 0.46 \div 10 = 0.04\dfrac{6}{10} = 0.04\dfrac{3}{5}$

10. $0.05 + 1.9 = 1.95$

11. $3.67 - 0.08 = 3.59$

12. $0.006 + 4.2 = 4.206$

13. done

14. $\left(\dfrac{1}{2}\right)(9 \text{ in})(12 \text{ in}) = 54 \text{ in}^2$

15. $\left(\dfrac{1}{2}\right)(3.08 \text{ m})(0.7 \text{ m}) = 1.078 \text{ m}^2$

16. $\$355 \times 0.20 = \$71.00$

17. $2{,}400 \text{ g} \times \dfrac{1 \text{ kg}}{1{,}000 \text{ g}} = 2.4 \text{ kg}$

$25 \text{ kg} - 2.4 \text{ kg} = 22.6 \text{ kg}$

18. $7H = 37.1 \text{ ft}^2$

$H = 37.1 \text{ ft}^2 \div 7 \text{ ft} = 5.3 \text{ ft}$

Systematic Review 21E

1.
$$6.844 \approx 6.84$$

$$9\overline{)61.600}$$
$$\underline{54\,000}$$
$$7\,600$$
$$\underline{7\,200}$$
$$400$$
$$\underline{360}$$
$$40$$
$$\underline{36}$$

2.
$$.018 \approx 0.02$$

$$20\overline{).370}$$
$$\underline{200}$$
$$170$$
$$\underline{160}$$

3.
$$.133 \approx 0.13$$

$$3\overline{).400}$$
$$\underline{300}$$
$$100$$
$$\underline{90}$$
$$10$$
$$\underline{9}$$

4.
$$.266 \text{ or } 0.2\overline{6}$$

$$6\overline{)1.600}$$
$$\underline{1\,200}$$
$$400$$
$$\underline{360}$$
$$40$$
$$\underline{36}$$

5.
$$23.33 \text{ or } 23.\overline{3}$$

$$3\overline{)70.00}$$
$$\underline{60\,00}$$
$$10\,00$$
$$\underline{9\,00}$$
$$1\,00$$
$$\underline{90}$$

6.
$$.10909 \text{ or } 0.10\overline{9}$$

$$11\overline{)1.20000}$$
$$\underline{1\,10000}$$
$$10000$$
$$\underline{9900}$$
$$100$$
$$\underline{99}$$

7. $12X = 0.28$

$X = 0.28 \div 12 = 0.02\frac{4}{12} = 0.02\frac{1}{3}$

8. $0.9Y = 0.5$

$Y = 0.5 \div 0.9 = 0.55\frac{5}{9}$

9. $1.3F = 5.4$

$F = 5.4 \div 1.3 = 4.15\frac{5}{13}$

10. $6.15 - 0.06 = 6.09$

11. $14.003 + 0.5 = 14.503$

12. $8.3 - 0.67 = 7.63$

13. $\frac{1}{2}(4\,\text{ft})(6\,\text{ft}) = 12\,\text{ft}^2$

14. $\frac{1}{2}(12\,\text{in})(16\,\text{in}) = 96\,\text{in}^2$

15. $\frac{1}{2}(1.02\,\text{m})(0.5\,\text{m}) = 0.255\,\text{m}^2$

16. $\$11.00 \times 1.06 = \$11.66$

17. $\$0.96 \div 12\,\text{oz} = \$0.08$ per oz

18. $\$0.56 \div 8\,\text{oz} = \$0.07$ per oz
The 8-ounce can is a better buy.

19. $4{,}180\,\text{m} \div \dfrac{1\,\text{km}}{1{,}000\,\text{m}} = 4.18\,\text{km}$

$4.18\,\text{km} - 4\,\text{km} = 0.18\,\text{km}$

20. $20H = 40.22\,\text{in}^2$

$H = 40.22\,\text{in}^2 \div 20\,\text{in}$

$H = 2.011\,\text{in}$ (rounds to 2.01 in)

Systematic Review 21F

1.
$$8.062 \approx 8.06$$

$$8\overline{)64.500}$$
$$\underline{64\,000}$$
$$500$$
$$\underline{480}$$
$$20$$
$$\underline{16}$$

2.
$$.304 \approx 0.30$$

$$15\overline{)4.570}$$
$$\underline{4\,500}$$
$$70$$
$$\underline{60}$$

3.
$$\begin{array}{r} 2.128 \approx 2.13 \\ 7\overline{)14.900} \\ \underline{14\ 000} \\ 900 \\ \underline{700} \\ 200 \\ \underline{140} \\ 60 \\ \underline{56} \end{array}$$

4.
$$\begin{array}{r} 2.242 \text{ or } 2.\overline{24} \\ 33\overline{)74.000} \\ \underline{66\ 000} \\ 8\ 000 \\ \underline{6\ 600} \\ 1\ 400 \\ \underline{1\ 320} \\ 80 \\ \underline{66} \end{array}$$

5.
$$\begin{array}{r} 50.909 \text{ or } 50.9\overline{0} \\ 11\overline{)560.000} \\ \underline{550\ 000} \\ 10\ 000 \\ \underline{9\ 900} \\ 100 \\ \underline{99} \end{array}$$

6.
$$\begin{array}{r} 1.11 \text{ or } 1.\overline{1} \\ 9\overline{)10.00} \\ \underline{9\ 00} \\ 1\ 00 \\ \underline{90} \\ 10 \\ \underline{9} \end{array}$$

7. $40X = 0.6$

   $X = 0.6 \div 40 = 0.01\frac{1}{2}$

8. $0.2Y = 0.077$

   $Y = 0.077 \div 0.2 = 0.38\frac{1}{2}$

9. $4.1F = 8.24$

   $F = 8.24 \div 4.1 = 2.00\frac{40}{41}$

10. $11 - 0.99 = 10.01$

11. $5.003 + 21 = 26.003$

12. $100 - 0.01 = 99.99$

13. $\frac{1}{2}(10 \text{ ft})(12 \text{ ft}) = 60 \text{ ft}^2$

14. $\frac{1}{2}(7 \text{ in})(24 \text{ in}) = 84 \text{ in}^2$

15. $\frac{1}{2}(1.3 \text{ m})(4.2 \text{ m}) = 2.73 \text{ m}^2$

16. $\$45.00 \times 1.08 = \$48.60$

17. $\$100.00 \times 3.15 = \$315.00$

18. $3.14(7 \text{ in})^2 = 154 \text{ in}^2$ (rounded)

    $\$10.00 \div 154 \text{ in}^2 = \$0.06$ per in^2

19. $3.14(6 \text{ in})^2 \approx 113 \text{ in}^2$

    $\$8.00 \div 113 \text{ in}^2 = \$0.07$ per in^2

    14 in is the better buy.

20. $4.25 \text{ cm} \times \dfrac{10 \text{ mm}}{1 \text{ cm}} = 42.5 \text{ mm}$

    $425 \text{ mm} - 42.5 \text{ mm} = 382.5 \text{ mm}$

Lesson Practice 22A

1. done

2. done

3. $0.09X + 4 = 4.9$

   $0.09X = 4.9 - 4$

   $0.09X = 0.9$

   $X = 0.9 \div 0.09$

   $X = 10$

4. $0.09(10) + 4 = 4.9$

   $0.9 + 4 = 4.9$

   $4.9 = 4.9$

5. $0.6A + 2.8 = 4.66$

   $0.6A = 4.66 - 2.8$

   $0.6A = 1.86$

   $A = 1.86 \div 0.6$

   $A = 3.1$

6. $0.6(3.1) + 2.8 = 4.66$

   $1.86 + 2.8 = 4.66$

   $4.66 = 4.66$

7. $4.6Q + 0.19 = 55.39$

   $4.6Q = 55.39 - 0.19$

   $4.6Q = 55.2$

   $Q = 55.2 \div 4.6$

   $Q = 12$

8. $4.6(12)+0.19=55.39$
$55.2+0.19=55.39$
$55.39=55.39$

9. $0.2M+\$10.50=\$15.50$
$0.2M=\$15.50-\$10.50$
$0.2M=\$5.00$
$M=\$5.00\div0.2$
$M=\$25.00$

10. $0.10H+5\text{ ft}=13\text{ ft}$
$0.10H=13\text{ ft}-5\text{ ft}$
$0.10H=8\text{ ft}$
$H=8\text{ ft}\div0.10=80\text{ ft}$

8. $2.2(0.6)+0.47=1.79$
$1.32+0.47=1.79$
$1.79=1.79$

9. $0.80M+\$14.40=\$254.40$
$0.80M=\$254.40-\$14.40$
$0.80M=\$240.00$
$M=\$240.00\div0.80$
$M=\$300.00$

10. $0.10M+\$10.00=\$15.00$
$0.10M=\$15.00-\$10.00$
$0.10M=\$5.00$
$M=\$5.00\div0.10$
$M=\$50.00$

Lesson Practice 22B

1. $1.1W+2.1=5.4$
$1.1W=5.4-2.1$
$1.1W=3.3$
$W=3.3\div1.1$
$W=3$

2. $1.1(3)+2.1=5.4$
$3.3+2.1=5.4$
$5.4=5.4$

3. $0.8D+0.2=0.76$
$0.8D=0.76-0.2$
$0.8D=0.56$
$D=0.56\div0.8$
$D=0.7$

4. $0.8(0.7)+0.2=0.76$
$0.56+0.2=0.76$
$0.76=0.76$

5. $0.13X+0.07=0.551$
$0.13X=0.551-0.07$
$0.13X=0.481$
$X=0.481\div0.13$
$X=3.7$

6. $0.13(3.7)+0.07=0.551$
$0.481+0.07=0.551$
$0.551=0.551$

7. $2.2R+0.47=1.79$
$2.2R=1.79-0.47$
$2.2R=1.32$
$R=1.32\div2.2$
$R=0.6$

Lesson Practice 22C

1. $0.7G+5.1=5.59$
$0.7G=5.59-5.1$
$0.7G=0.49$
$G=0.49\div0.7$
$G=0.7$

2. $0.7(0.7)+5.1=5.59$
$0.49+5.1=5.59$
$5.59=5.59$

3. $3.4X+0.01=0.622$
$3.4X=0.622-0.01$
$3.4X=0.612$
$X=0.612\div3.4$
$X=0.18$

4. $3.4(0.18)+0.01=0.622$
$0.612+0.01=0.622$
$0.622=0.622$

5. $8.7Y+1.1=5.45$
$8.7Y=5.45-1.1$
$8.7Y=4.35$
$Y=4.35\div8.7$
$Y=0.5$

6. $8.7(0.5)+1.1=5.45$
$4.35+1.1=5.45$
$5.45=5.45$

7. $0.01B+3.3=3.341$
$0.01B=3.341-3.3$
$0.01B=0.041$
$B=0.041\div0.01$
$B=4.1$

8. $0.01(4.1) + 3.3 = 3.341$
$0.041 + 3.3 = 3.341$
$3.341 = 3.341$

9. $0.5P + 1\ \text{gal} = 4\ \text{gal}$
$0.5P = 4\ \text{gal} - 1\ \text{gal}$
$0.5P = 3\ \text{gal}$
$P = 3\ \text{gal} \div 0.5$
$P = 6\ \text{gal for job}$
$6\ \text{gal} - 4\ \text{gal} = 2\ \text{gal to buy}$

10. $2.1R + 0.6" = 32.1"$
$2.1R = 32.1" - 0.6"$
$2.1R = 31.5"$
$R = 31.5" \div 2.1$
$R = 15"$

Systematic Review 22D

1. $6.8X + 0.9 = 22.66$
$6.8X = 22.66 - 0.9$
$6.8X = 21.76$
$X = 21.76 \div 6.8$
$X = 3.2$

2. $6.8(3.2) + 0.9 = 22.66$
$21.76 + 0.9 = 22.66$
$22.66 = 22.66$

3. $5.1Q + 4 = 4.306$
$5.1Q = 4.306 - 4$
$5.1Q = 0.306$
$Q = 0.306 \div 5.1$
$Q = 0.06$

4. $5.1(0.06) + 4 = 4.306$
$0.306 + 4 = 4.306$
$4.306 = 4.306$

5.
$$8.908 \approx 8.91$$
$$6\overline{)53.450}$$
$$\underline{48\ 000}$$
$$5\ 450$$
$$\underline{5\ 400}$$
$$50$$
$$\underline{48}$$

6.
$$.271 \approx 0.27$$
$$7\overline{)1.900}$$
$$\underline{1\ 400}$$
$$500$$
$$\underline{490}$$
$$10$$
$$\underline{7}$$

7.
$$.204 \approx 0.20$$
$$14\overline{)2.860}$$
$$\underline{2\ 800}$$
$$60$$
$$\underline{56}$$

8. $0.382\ \text{hm}$

9. $0.00017\ \text{L}$

10. $5\ \text{dag}$

11. done

12. $10\text{m} \times 10\text{m} \times 10\text{m} = 1,000\ \text{m}^3$

13. $1.2\ \text{ft} \times 1.2\ \text{ft} \times 1.2\ \text{ft} = 1.728\ \text{ft}^3$

14. $1.7\text{m} + 0.8\text{m} + 1.7\text{m} + 0.8\text{m} = 5\ \text{m}$

15. $11\ \text{in} + 3\ \text{in} + 11\ \text{in} + 3\ \text{in} = 28\ \text{in}$

16. $3.14(6\ \text{ft}) = 18.84\ \text{ft}$

17. $\$16.64 \times 1.15 = \$19.136$
($19.14 rounded)

18. $0.5L + 12\ \text{ft}^2 = 132\ \text{ft}^2$
$0.5L = 132\ \text{ft}^2 - 12\ \text{ft}^2$
$0.5L = 120\ \text{ft}^2$
$L = 120\ \text{ft}^2 \div 0.5$
$L = 240\ \text{ft}^2$

Systematic Review 22E

1. $0.4T + 3.6 = 12$
$0.4T = 12 - 3.6$
$0.4T = 8.4$
$T = 8.4 \div 0.4$
$T = 21$

2. $0.4(21) + 3.6 = 12$
$8.4 + 3.6 = 12$
$12 = 12$

3. $9.9W + 0.7 = 5.749$
$9.9W = 5.749 - 0.7$
$9.9W = 5.049$
$W = 5.049 \div 9.9$
$W = 0.51$

4. $9.9(0.51) + 0.7 = 5.749$
$5.049 + 0.7 = 5.749$
$5.749 = 5.749$

5. 1.833 or $1.8\overline{3}$

$$24 \overline{)44.000}$$
$\underline{24\ 000}$
$19\ 200$
$\underline{800}$
720
$\underline{80}$
72

6. 351.66 or $351.\overline{6}$

$$6 \overline{)2110.00}$$
$\underline{1800\ 00}$
$310\ 00$
$\underline{300\ 00}$
$10\ 00$
$\underline{600}$
400
$\underline{360}$
40
$\underline{36}$
4

7. 9.44 or $9.\overline{4}$

$$9 \overline{)85.00}$$
$\underline{81\ 00}$
$4\ 00$
$\underline{3\ 60}$
40
$\underline{36}$

8. 0.75 kg

9. 24.5 cm

10. 60 hl

11. $0.03 \text{ ft} \times 0.03 \text{ ft} \times 0.03 \text{ ft} = 0.000027$

12. $8m \times 8m \times 8m = 512 \text{ m}^3$

13. $6.4 \text{ in} \times 6.4 \text{ in} \times 6.4 \text{ in} = 262.144 \text{ in}^3$

14. $1.7m \times 0.8m = 1.36 \text{ m}^2$

15. $11 \text{ in} \times 2 \text{ in} = 22 \text{ in}^2$

16. $3.14(3 \text{ ft})^2 = 28.26 \text{ ft}^2$

17. $\$48.00 \times 0.55 = \$26.40$
$\$48.00 - \$26.40 = \$21.60$
$\$21.60 \times 1.06 = \$22.90$

18. $50,000 \text{ g} = 50 \text{ kg}$

$50 \text{ kg} \times \dfrac{2.2 \text{ lb}}{1 \text{ kg}} = 110 \text{ lb}$

19. $\dfrac{5}{9} \times 54$ games

$54 \div 9 = 6$

$6 \times 5 = 30$ games won

20. $0.7M + \$5.00 = \$26.49$
$0.7M = \$26.49 - \$5.00$
$0.7M = \$21.49$
$M = \$21.49 \div 0.7 = \$30.70$ needed

$\$30.70 - \$26.49 = \$4.21$ still to go

Systematic Review 22F

1. $0.08G + 0.59 = 1.39$
$0.08G = 1.39 - 0.59$
$0.08G = 0.8$
$G = 0.8 \div 0.08$
$G = 10$

2. $0.08(10) + 0.59 = 1.39$
$0.8 + 0.59 = 1.39$
$1.39 = 1.39$

3. $0.3X + 2.4 = 2.58$
$0.3X = 2.58 - 2.4$
$0.3X = 0.18$
$X = 0.18 \div 0.3$
$X = 0.6$

4. $0.3(0.6) + 2.4 = 2.58$
$0.18 + 2.4 = 2.58$
$2.58 = 2.58$

5. $.72\frac{6}{7}$

$$7 \overline{)5.10}$$
$\underline{4\ 90}$
20
$\underline{14}$

6. $6.27\frac{1}{2}$

$$2 \overline{)12.55}$$
$\underline{12\ 00}$
55
$\underline{40}$
15
$\underline{14}$

7.
$$8\overline{)12.50} \quad 1.56\frac{2}{8} = 1.56\frac{1}{4}$$
$$\underline{8\ 00}$$
$$4\ 50$$
$$\underline{4\ 00}$$
$$50$$
$$\underline{48}$$

8. 6.5 g

9. 0.321 km

10. 62 ml

11. $\frac{5}{2}$ ft $\times \frac{5}{2}$ ft $\times \frac{5}{2}$ ft $= \frac{125}{8}$ ft $= 15\frac{5}{8}$ ft^3

12. $0.9\text{m} \times 0.9\text{m} \times 0.9\text{m} = 0.729 \text{ m}^3$

13. $4.8 \text{ in} \times 4.8 \text{ in} \times 4.8 \text{ in} = 110.592 \text{ in}^3$

14. $\frac{1}{2}(6 \text{ ft})(16 \text{ ft}) = 48 \text{ ft}^2$

15. $10 \text{ ft} + 10 \text{ ft} + 16 \text{ ft} = 36 \text{ ft}$

16. $\frac{1}{2}(3 \text{ m})(5.1 \text{ m}) = 7.65 \text{ m}^2$

17. $4.5\text{m} + 5.1\text{m} + 3.3\text{m} = 12.9 \text{ m}$

18. $3 \text{ ft} \times 3 \text{ ft} = 9 \text{ ft}^2$

19. $12 \text{ ft} \times 15 \text{ ft} = 180 \text{ ft}^2$

$$180 \text{ ft}^2 \div \frac{9 \text{ ft}^2}{1 \text{ yd}^2} = 20 \text{ yd}^2$$

20. $20 \text{ yd}^2 \times (\$9.99/\text{yd}^2) = \$199.80$

Lesson Practice 23A

1. done

2. $6 \div 7 = 0.85\frac{5}{7} = 85\frac{5}{7}\%$

3. $3 \div 10 = 0.30 = 30\%$

4. $2 \div 5 = 0.40 = 40\%$

5. $7 \div 8 = 0.87\frac{1}{2} = 87\frac{1}{2}\%$

6. $10 \div 11 = 0.90\frac{10}{11} = 90\frac{10}{11}\%$

7. $\frac{13}{15} = 13 \div 15 = 0.86\frac{2}{3} = 86\frac{2}{3}\%$

8. $\frac{1}{6} = 1 \div 6 = 0.16\frac{2}{3} = 16\frac{2}{3}\%$

9. $\frac{16}{50} = 16 \div 50 = 0.32 = 32\%$

10. $50 - 16 = 34$ not wearing hats

$$\frac{34}{50} = 34 \div 50 = 0.68 = 68\%$$

(or $100\% - 32\% = 68\%$)

Lesson Practice 23B

1. $7 \div 10 = 0.70 = 70\%$

2. $1 \div 2 = 0.50 = 50\%$

3. $4 \div 9 = 0.44\frac{4}{9} = 44\frac{4}{9}\%$

4. $3 \div 11 = 0.27\frac{3}{11} = 27\frac{3}{11}\%$

5. $3 \div 5 = 0.60 = 60\%$

6. $1 \div 12 = 0.08\frac{1}{3} = 8\frac{1}{3}\%$

7. $3 \div 4 = 0.75 = 75\%$

8. $5 \div 6 = 0.83\frac{1}{3} = 83\frac{1}{3}\%$

9. $25 - 3 = 22$ correct

$22 \div 25 = 0.88 = 88\%$

10. $35 \div 100 = 0.35 = 35\%$

Lesson Practice 23C

1. $1 \div 7 = 0.14\frac{2}{7} = 14\frac{2}{7}\%$

2. $2 \div 9 = 0.22\frac{2}{9} = 22\frac{2}{9}\%$

3. $4 \div 5 = 0.80 = 80\%$

4. $5 \div 12 = 0.41\frac{2}{3} = 41\frac{2}{3}\%$

5. $2 \div 3 = 0.66\frac{2}{3} = 66\frac{2}{3}\%$

6. $3 \div 8 = 0.37\frac{1}{2} = 37\frac{1}{2}\%$

7. $87 \div 100 = 0.87 = 87\%$

8. $5 \div 8 - 0.62\frac{1}{2} = 62\frac{1}{2}\%$

9. $1 \div 3 = 0.33\frac{1}{3} = 33\frac{1}{3}\%$

10. $1 \div 2 = 0.50 = 50\%$

Systematic Review 23D

1. $1 \div 4 = 0.25 = 25\%$

2. $7 \div 9 = 0.77\frac{7}{9} = 77\frac{7}{9}\%$

3. $5 \div 16 = 0.31\frac{1}{4} = 31\frac{1}{4}\%$

4. $1 \div 8 = 0.12\frac{1}{2} = 12\frac{1}{2}\%$

5. $0.17X + 1.2 = 1.234$
$0.17X = 1.234 - 1.2$
$0.17X = 0.034$
$X = 0.034 \div 0.17$
$X = 0.2$

6. $0.17(.2) + 1.2 = 1.234$
$0.034 + 1.2 = 1.234$
$1.234 = 1.234$

7.
$$\begin{array}{r} 2.046 \approx 2.05 \\ 13\overline{)26.600} \\ \underline{26\ 000} \\ 600 \\ \underline{520} \\ 80 \\ \underline{78} \end{array}$$

8.
$$\begin{array}{r} .822 \approx 0.82 \\ 9\overline{)7.400} \\ \underline{7\ 200} \\ 200 \\ \underline{180} \\ 20 \\ \underline{18} \end{array}$$

9.
$$\begin{array}{r} .198 \approx 0.20 \\ 23\overline{)4.570} \\ \underline{2\ 300} \\ 2\ 270 \\ \underline{2\ 070} \\ 200 \\ \underline{184} \end{array}$$

10. done

11. $12m \times 4m \times 4.1m = 196.8 \ m^3$

12. $2.3 \ ft \times 1.8 \ ft \times 3.2 \ ft = 13.248 \ ft^3$

13. $38 \div 45 = 0.84\frac{4}{9} = 84\frac{4}{9}\%$

14. $6 \div 8 = 0.75 = 75\%$

15. $4mi \times 2mi \times 2mi = 16 \ mi^3$

16. $12in \times 12in = 144 \ in^2$

17. $10 \ ft \times 10 \ ft = 100 \ ft^2$
$100 \ ft^2 \times 144 \ in^2/ft^2 = 14,400 \ in^2$

18. $\$15.99 \times 0.40 = \$6.396$ or
$\$6.40$ off the regular price

Systematic Review 23E

1. $1 \div 9 = 0.11\frac{1}{9} = 11\frac{1}{9}\%$

2. $2 \div 3 = 0.66\frac{2}{3} = 66\frac{2}{3}\%$

3. $11 \div 14 = 0.78\frac{4}{7} = 78\frac{4}{7}\%$

4. $4 \div 7 = 0.57\frac{1}{7} = 57\frac{1}{7}\%$

5. $0.3Y + 0.09 = 1.44$
$0.3Y = 1.44 - 0.09$
$0.3Y = 1.35$
$Y = 1.35 \div 0.3$
$Y = 4.5$

6. $0.3(4.5) + 0.09 = 1.44$
$1.35 + 0.09 = 1.44$
$1.44 = 1.44$

7.
$$\begin{array}{r} 8.909 \ \text{or} \ 8.\overline{90} \\ 11\overline{)98.000} \\ \underline{88\ 000} \\ 10\ 000 \\ \underline{9\ 900} \\ 100 \\ \underline{99} \end{array}$$

8.
$$\begin{array}{r} 12.9833 \ \text{or} \ 12.98\overline{3} \\ 6\overline{)77.9000} \\ \underline{60\ 0000} \\ 17\ 0000 \\ \underline{12\ 0000} \\ 5\ 0000 \\ \underline{5000} \\ 4800 \\ \underline{200} \\ 180 \end{array}$$

9.

$$104.166 \text{ or } 104.1\overline{6}$$

$$48\overline{)5000.000}$$
$$\underline{4800\ 000}$$
$$200\ 000$$
$$\underline{192\ 000}$$
$$8\ 000$$
$$\underline{4\ 800}$$
$$3\ 200$$
$$\underline{2\ 880}$$
$$320$$
$$\underline{288}$$

10. $7\text{in} \times 6\text{in} \times 5\text{in} = 210 \text{ in}^3$

11. $7.5\text{m} \times 2.4\text{m} \times 2\text{m} = 36 \text{ m}^3$

12. $\dfrac{7}{2}\text{ft} \times \dfrac{9}{4}\text{ft} \times \dfrac{29}{8}\text{ft} = \dfrac{1827}{64} = 28\dfrac{35}{64} \text{ ft}^3$

13. $102 + 60 = 162$ total games

$$102 \div 162 = 0.62\dfrac{26}{27} = 62\dfrac{26}{27}\%$$

14. $60 \div 162 = 0.37\dfrac{1}{27} = 37\dfrac{1}{27}\%$

$$0.37\dfrac{1}{27} + 62\dfrac{26}{27} = 0.99\dfrac{27}{27} = 100\%$$

15. $\$47.50 \times 1.08 = \$51.30$

16. round pizza:

$$3.14(6 \text{ in})^2 = 113 \text{ in}^2$$

$$\$5.65 \div 113 \text{ in}^2 = \$0.05 \text{ per in}^2$$

rectangular pizza:

$$12 \text{ in} \times 24 \text{ in} = 288 \text{ in}^2$$

$$\$11.52 \div 288 \text{ in}^2 = \$0.04 \text{ per in}^2$$

Rectangular pizza has better price.

17. $288 \text{ in}^2 \times \dfrac{1 \text{ ft}^2}{144 \text{ in}^2} = 2 \text{ ft}^2$

or $12 \text{ in} \times 24 \text{ in} = 1 \text{ ft} \times 2 \text{ ft} = 2 \text{ ft}^2$

18. $\$4.20 \times 0.75 = \$3.15$

19. $36\text{in} + 42\text{in} + 36\text{in} + 42\text{in} = 156 \text{ in}$

20. $156 \text{ ft} \div 1 \text{ ft}/12 \text{ in} = 13 \text{ ft}$

Systematic Review 23F

1. $7 \div 15 = 0.46\dfrac{2}{3} = 46\dfrac{2}{3}\%$

2. $3 \div 7 = 0.42\dfrac{6}{7} = 42\dfrac{6}{7}\%$

3. $5 \div 9 = 0.55\dfrac{5}{9} = 55\dfrac{5}{9}\%$

4. $1 \div 10 = 0.10 = 10\%$

5.
$$6.7\text{R} + 0.4 = 6.43$$
$$6.7\text{R} = 6.43 - 0.4$$
$$6.7\text{R} = 6.03$$
$$\text{R} = 6.03 \div 6.7$$
$$\text{R} = 0.9$$

6.
$$6.7(0.9) + 0.4 = 6.43$$
$$6.03 + 0.4 = 6.43$$
$$6.43 = 6.43$$

7.

$$1.56\dfrac{2}{33}$$
$$33.\overline{)51.50}$$
$$\underline{3300}$$
$$1850$$
$$\underline{1650}$$
$$200$$
$$\underline{198}$$
$$2$$

8.

$$8.23\dfrac{6}{8} = 8.23\dfrac{3}{4}$$
$$8.\overline{)65.90}$$
$$\underline{6400}$$
$$190$$
$$\underline{160}$$
$$30$$
$$\underline{24}$$
$$6$$

9.

$$11.96\dfrac{4}{6} = 11.96\dfrac{2}{3}$$
$$06.\overline{)71.80}$$
$$\underline{6000}$$
$$1180$$
$$\underline{600}$$
$$580$$
$$\underline{540}$$
$$40$$
$$\underline{36}$$
$$4$$

10. $80\text{in} \times 90\text{in} \times 60\text{in} = 432,000 \text{ in}^3$

11. $13.1\text{m} \times 4.6\text{m} \times 2.9\text{m} = 174.754 \text{ m}^3$

12. $0.09\text{ft} \times 0.05\text{ft} \times 1\text{ft} = 0.0045 \text{ ft}^3$

13. $3\text{ft} \times 3\text{ft} \times 3\text{ft} = 27 \text{ ft}^3$

14. $27 \times 2\dfrac{1}{3} = \dfrac{\overset{9}{\cancel{27}}}{1} \times \dfrac{7}{\cancel{3}} = \dfrac{63}{1} = 63 \text{ ft}^3$

15. $6 \text{ yd}^3 \times 27 \text{ ft}^3 / \text{yd}^3 = 162 \text{ ft}^3$

16. $6 \times \$55.50 = \$333$

17. $6\text{yd}^3 \div 8\text{yd}^3 = 0.75 = 75\%$

18. $34 \div 8 = 4.25$, so 5 trips

19. $\$32 \times 0.30 = \$9.60$
    $\$32 - \$9.60 = \$22.40$ before tax
    $\$22.40 \times 1.04 = \$23.296$ or $\$23.30$
    $\$25 - \$23.30 = \$1.70$

20. $7.8H = 27.3$

    $\dfrac{\cancel{7.8}}{\cancel{7.8}} H = \dfrac{27.3}{7.8}$

    $27.3 \text{ ft}^2 \div 7.8 \text{ ft} = 3.5 \text{ ft}$

Lesson Practice 24A

1. done

2. $\dfrac{96}{100} = \dfrac{24}{25}$

3. $\dfrac{982}{1000} = \dfrac{491}{500}$

4. $\dfrac{6}{1000} = \dfrac{3}{500}$

5. $\dfrac{18}{1000} = \dfrac{9}{500}$

6. $\dfrac{9}{10}$

7. $\dfrac{885}{1000} = \dfrac{177}{200}$

8. $\dfrac{84}{100} = \dfrac{21}{25}$

9. $\dfrac{32}{100} = \dfrac{8}{25}$

10. $\dfrac{15}{1000} = \dfrac{3}{200}$

11. $\dfrac{20}{100} = \dfrac{1}{5}$; 5 pieces

12. $\dfrac{45}{100} = \dfrac{9}{20}$; 9 people

Lesson Practice 24B

1. $\dfrac{915}{1000} = \dfrac{183}{200}$

2. $\dfrac{53}{1000}$

3. $\dfrac{7}{1000}$

4. $\dfrac{62}{100} = \dfrac{31}{50}$

5. $\dfrac{75}{100} = \dfrac{3}{4}$

6. $\dfrac{52}{100} = \dfrac{13}{25}$

7. $\dfrac{28}{1000} = \dfrac{7}{250}$

8. $\dfrac{725}{1000} = \dfrac{29}{40}$

9. $\dfrac{6}{10} = \dfrac{3}{5}$

10. $\dfrac{95}{100} = \dfrac{19}{20}$

11. $\dfrac{4}{10} = \dfrac{2}{5}$; 2 juncos

12. $\dfrac{125}{1000} = \dfrac{1}{8}$; 8 pieces

Lesson Practice 24C

1. $\dfrac{68}{100} = \dfrac{17}{25}$

2. $\dfrac{25}{1000} = \dfrac{1}{40}$

3. $\dfrac{84}{100} = \dfrac{21}{25}$

4. $\dfrac{16}{1000} = \dfrac{2}{125}$

5. $\dfrac{8}{1000} = \dfrac{1}{125}$

6. $\dfrac{7}{10}$

7. $\dfrac{325}{1000} = \dfrac{13}{40}$

8. $\dfrac{743}{1000}$

9. $\dfrac{56}{100} = \dfrac{14}{25}$

10. $\dfrac{11}{100}$

11. $\dfrac{4}{100} = \dfrac{1}{25}$; 25 pieces

12. $0.75 \times 4 = 3$ houses

Systematic Review 24D

1. $\dfrac{25}{1000} = \dfrac{1}{40}$

2. $\dfrac{18}{100} = \dfrac{9}{50}$

3. $1 \div 5 = 0.20 = 20\%$

4. $5 \div 8 = 0.62\dfrac{1}{2} = 62\dfrac{1}{2}\%$

5. $0.19X + 0.73 = 0.825$
$\qquad 0.19X = 0.825 - 0.73$
$\qquad 0.19X = 0.095$
$\qquad\quad X = 0.095 \div 0.19$
$\qquad\quad X = 0.5$

6. $0.19(0.5) + 0.73 = 0.825$
$\qquad 0.095 + 0.73 = 0.825$
$\qquad\qquad 0.825 = 0.825$

7.
$$\begin{array}{r} 2.306 = 2.31 \\ 15\overline{)34.600} \\ \underline{30000} \\ 4600 \\ \underline{4500} \\ 100 \\ \underline{90} \\ 10 \end{array}$$

8.
$$\begin{array}{r} .837 = 0.84 \\ 8.\overline{)6.700} \\ \underline{6400} \\ 300 \\ \underline{240} \\ 60 \\ \underline{56} \\ 4 \end{array}$$

9.
$$\begin{array}{r} .036 = 0.04 \\ 34.\overline{)1.230} \\ \underline{1020} \\ 210 \\ \underline{204} \\ 6 \end{array}$$

10. $2 + 7 + 9 = 18$
$18 \div 3 = 6$

11. $5 + 5 + 9 + 13 = 32$
$32 \div 4 = 8$

12. $2 + 5 + 8 + 9 = 24$
$24 \div 4 = 6$

13. $77 + 80 + 95 + 100 = 352$
$352 \div 4 = 88$

14. $19.06 \text{ ft/sec} \times 60 \text{ sec} = 1{,}143.6 \text{ ft}$

15. $1.15 \times 5 \text{ hr} = 5.75 \text{ hr}$

16. $\dfrac{1}{2}(0.24\text{m} \times 0.5\text{m}) = 0.06 \text{ m}^2$

17. $2{,}500 \text{ ml} \times \dfrac{1 \text{ L}}{1{,}000 \text{ ml}} = 2.5 \text{ liters}$

18. $4 \text{ km} \times 10 \text{ hm/km} = 40 \text{ hm}$

Systematic Review 24E

1. $\dfrac{625}{1000} = \dfrac{5}{8}$

2. $\dfrac{21}{100}$

3. $2 \div 7 = 0.28\dfrac{4}{7} = 28\dfrac{4}{7}\%$

4. $1 \div 4 - 0.25 = 25\%$

5. $0.3X + 9.1 = 9.244$
$\qquad 0.3X = 9.244 - 9.1$
$\qquad 0.3X = 0.144$
$\qquad\quad X = 0.144 \div 0.3$
$\qquad\quad X = 0.48$

6. $0.3(0.48) + 9.1 = 9.244$
$\qquad 0.144 + 9.1 = 9.244$
$\qquad\qquad 9.244 = 9.244$

7.
$$\begin{array}{r} 9.044 = 9.0\overline{4} \\ 9.\overline{)81.400} \\ \underline{81000} \\ 400 \\ \underline{360} \\ 40 \\ \underline{36} \\ 4 \end{array}$$

8.
$$\begin{array}{r} 48.66 = 48.\overline{6} \\ 3.\overline{)146.00} \\ \underline{12000} \\ 2600 \\ \underline{2400} \\ 200 \\ \underline{180} \\ 20 \\ \underline{18} \\ 2 \end{array}$$

9.
$$
\begin{array}{r}
4.8181 = 4.\overline{81} \\
11\overline{)53.0000} \\
\underline{440000} \\
90000 \\
\underline{88000} \\
2000 \\
\underline{1100} \\
900 \\
\underline{880} \\
20
\end{array}
$$

10. $4+5+6=15$

$15 \div 3 = 5$

11. $5+8+8=21$

$21 \div 3 = 7$

12. $6+7+9+10=32$

$32 \div 4 = 8$

13. $76+81+89+92=338$

$338 \div 4 = 84.5$

14. $10m \times 10m \times 10m = 1,000 \ m^3$

15. $6 \ in + 9 \ in + 11 \ in = 26 \ in$

16. $10,000,000 \ m \times \dfrac{1 \ km}{1,000 \ m} = 10,000 \ km$

17. $384 \ oz \times 0.75 = 288 \ oz$

18. $288 \ oz \times \dfrac{1 \ lb}{16 \ oz} = 18 \ lb$

19. $\$9.00 \times 1.10 = \$9.90$

20. $3.14(4 \ in)^2 = 50.24 \ in^2$

Systematic Review 24F

1. $\dfrac{92}{100} = \dfrac{23}{25}$

2. $\dfrac{45}{1000} = \dfrac{9}{200}$

3. $1 \div 9 = 0.11\dfrac{1}{9} = 11\dfrac{1}{9}\%$

4. $3 \div 5 = 0.60 = 60\%$

5. $0.2X + 75 = 75.18$

$0.2X = 75.18 - 75$

$0.2X = 0.18$

$X = 0.18 \div 0.2$

$X = 0.9$

6. $0.2(0.9) + 75 = 75.18$

$0.18 + 75 = 75.18$

$75.18 = 75.18$

7.
$$
\begin{array}{r}
166.66\dfrac{4}{6} = 166.66\dfrac{2}{3} \\
6.\overline{)1000.00} \\
\underline{60000} \\
40000 \\
\underline{36000} \\
4000 \\
\underline{3600} \\
400 \\
\underline{360} \\
40 \\
\underline{36} \\
4
\end{array}
$$

8.
$$
\begin{array}{r}
.32\dfrac{1}{2} \\
2.\overline{)0.65} \\
\underline{60} \\
5 \\
\underline{4} \\
1
\end{array}
$$

9.
$$
\begin{array}{r}
2.11\dfrac{3}{17} \\
17.\overline{)35.90} \\
\underline{3400} \\
190 \\
\underline{170} \\
20 \\
\underline{17} \\
3
\end{array}
$$

10. $3+6+9=18$

$18 \div 3 = 6$

11. $4+8+10+14=36$

$36 \div 4 = 9$

12. $1+3+6+10=20$

$20 \div 4 = 5$

13. $21 \ in + 34 \ in + 21 \ in + 34 \ in = 110 \ in$

14. $(2)(3.14)(4 \ ft) = 25.12 \ ft$

15. $\$2,500 \times 1.17 = \$2,925$

16. $250 \div 100 = 2.5$, so he must buy 3 packs

17. $\$1.29 \times 3 = \$3.87$

$\$3.87 \times 1.05 \approx \$4.06$

18. $3.25 \times 40 \ lb = 130 \ lb$

19. $130 \ lb \times \dfrac{1 \ bag}{10 \ lb} = 13 \ bags$

20. $13 \times \$1.45 = \$18.85$

Lesson Practice 25A

1. done

2. $\dfrac{3+8+8+8+9+9+11}{7} = 8$
   mean = 8
   median = 8
   mode = 8

3. $\dfrac{3+8+9+9+11}{5} = 8$
   mean = 8
   median = 9
   mode = 9

4. $\dfrac{3+4+6+6+6+8+9}{7} = 6$
   mean = 6
   median = 6
   mode = 6

5. $\dfrac{3+8+8+10+12+19}{6} = 10$
   mean = 10
   median = (8+10)/2 = 9
   mode = 8

6. $\dfrac{11+12+15+15+33+34}{6} = 20$
   mean = 20
   median = 15
   mode = 15

7. 6

8. 29° was the average temperature

9. 3

Lesson Practice 25B

1. $\dfrac{10+11+15+15+15+16+23}{7} = 15$
   mean = 15
   median = 15
   mode = 15

2. $\dfrac{1+1+1+1+2+2+2+3+3+4}{10} = 2$
   mean = 2
   median = 2
   mode = 1

3. $\dfrac{3+4+7+7+7+9+19}{7} = 8$
   mean = 8
   median = 7
   mode = 7

4. $\dfrac{13+14+24+24+25}{5} = 20$
   mean = 20
   median = 24
   mode = 24

5. $\dfrac{15+17+17+17+19+23}{6} = 18$
   mean = 18
   median = 17
   mode = 17

6. $\dfrac{9+10+10+10+10+11+13+13+13}{9} = 11$
   mean = 11
   median = 10
   mode = 10

7. 17°+30°+32°+35°+40° = 154°
   154°÷5 = 30.8°

8. 98

9. median

Lesson Practice 25C

1. $\dfrac{5+8+8+9+15}{5} = 9$
   mean = 9
   median = 8
   mode = 8

2. $\dfrac{9+11+11+14+20+23+24}{7} = 16$
   mean = 16
   median = 14
   mode = 11

3. $\dfrac{7+7+7+7+14+18}{6} = 10$
   mean = 10
   median = 7
   mode = 7

4. $\dfrac{1+6+6+19}{4} = 8$
   mean = 8
   median = 6
   mode = 6

5. $\dfrac{11+13+13+13+19+21}{6} = 15$

mean = 15
median = 13
mode = 13

6. $\dfrac{8+9+11+14+15+15+19}{7} = 13$

mean = 13
median = 14
mode = 15

7. $5+6+7+8+9 = 35$
$35 \div 5 = 7$

8. 5'10"; median

9. 8; mode

Systematic Review 25D

1. $\dfrac{5+6+6+9+10+10+10}{7} = 8$

mean = 8
median = 9
mode = 10

2. $\dfrac{7+8+11+11+13}{5} = 10$

mean = 10
median = 11
mode = 11

3. $\dfrac{13}{100}$

4. $\dfrac{350}{1000} = \dfrac{35}{100} = \dfrac{7}{20}$

5. $2 \div 9 = 0.22\dfrac{2}{9} = 22\dfrac{2}{9}\%$

6. $1 \div 10 = 0.10 = 10\%$

7. $0.7X + 1.5 = 3.88$
$0.7X = 3.88 - 1.5$
$0.7X = 2.38$
$X = 2.38 \div 0.7$
$X = 3.4$

8. $0.7(3.4) + 1.5 = 3.88$
$2.38 + 1.5 = 3.88$
$3.88 = 3.88$

9. $\pi(4.5 \text{ in})^2 \approx 63.59 \text{ in}^2$

10. $\pi(9 \text{ in}) \approx 28.26 \text{ in}$

11. $8 \times 8 = 64$

12. $3 \times 3 = 9$

13. $10 \times 10 = 100$

14. $15 \text{ ft} + 15 \text{ ft} + 24 \text{ ft} = 54 \text{ ft}$

15. $12 \text{ m} + 16 \text{ m} + 20 \text{ m} = 48 \text{ m}$

16. $1.5 \text{ ft} + 2.5 \text{ ft} + 2 \text{ ft} = 6 \text{ ft}$

17. $\dfrac{1}{2}(24 \text{ ft} \times 9 \text{ ft}) = 108 \text{ ft}^2$

18. $\dfrac{1}{2}(12 \text{ m} \times 16 \text{ m}) = 96 \text{ m}^2$

19. $\dfrac{1}{2}(1.2 \text{ ft} \times 2.5 \text{ ft}) = 1.5 \text{ ft}^2$

20. $50 \text{ g} \times 1{,}000 \text{ mg/g} = 50{,}000 \text{ mg total}$
$50{,}000 \text{ mg} \div 100 \text{ tablets} = 500 \text{ mg/tablet}$

Systematic Review 25E

1. $\dfrac{2+4+4+4+7+9}{6} = 5$

mean = 5
median = 4
mode = 4

2. $\dfrac{3+8+8+15+16+29+33}{7} = 16$

mean = 16
median = 15
mode = 8

3. $\dfrac{8}{10} = \dfrac{4}{5}$

4. $\dfrac{48}{100} = \dfrac{12}{25}$

5. $5 \div 6 = 0.83\dfrac{1}{3} = 83\dfrac{1}{3}\%$

6. $3 \div 15 = 0.2 = 20\%$

7. $1.1X + 2.2 = 9.9$
$1.1X = 9.9 - 2.2$
$1.1X = 7.7$
$X = 7.7 \div 1.1$
$X = 7$

8. $1.1(7) + 2.2 = 9.9$
$7.7 + 2.2 = 9.9$
$9.9 = 9.9$

9. $3.14(5 \text{ in})^2 = 78.5 \text{ in}^2$

10. $3.14(10 \text{ in}) = 31.4 \text{ in}$

11. $2 \times 2 \times 2 = 8$

12. $1 \times 1 \times 1 \times 1 \times 1 = 1$

13. $7 \times 7 = 49$

14. $2.6 \text{ ft} + 2.6 \text{ ft} + 2.6 \text{ ft} + 2.6 \text{ ft} = 10.4 \text{ ft}$

15. $0.1 \text{ m} + 0.51 \text{ m} + 0.1 \text{ m} + 0.51 \text{ m} = 1.22 \text{ m}$

16. $3.1 \text{ ft} + 5.4 \text{ ft} + 3.1 \text{ ft} + 5.4 \text{ ft} = 17 \text{ ft}$

17. $2.6 \text{ ft} \times 2.6 \text{ ft} = 6.76 \text{ ft}^2$

18. $0.1 \text{ m} \times 0.51 \text{ m} = 0.051 \text{ m}^2$

19. $3.1 \text{ ft} \times 5.4 \text{ ft} = 16.74 \text{ ft}^2$

20. $30,300 \text{ L} \div 1,000 \text{ L/kl} = 30.3 \text{ kl}$

18. $18 \text{ mm} \times 6 \text{ mm} = 108 \text{ mm}^2$

19. $2.4 \text{ in} \times 1.2 \text{ in} = 2.88 \text{ in}^2$

20. $3 \text{ dm} \times \dfrac{10 \text{ cm}}{1 \text{ dm}} = 30 \text{ one-cm pieces}$

Systematic Review 25F

1. $\dfrac{1+1+1+1+1+5+6+8}{8} = 3$

   mean = 3

   median = 1

   mode = 1

2. $\dfrac{5+9+9+9+13+15}{6} = 10$

   mean = 10

   median = 9

   mode = 9

3. $\dfrac{8}{100} = \dfrac{2}{25}$

4. $\dfrac{175}{1000} = \dfrac{7}{40}$

5. $9 \div 10 = 0.9 = 90\%$

6. $7 \div 8 = 0.87\frac{1}{2} = 87\frac{1}{2}\%$

7. $8X + 0.09 = 1.69$

   $\quad 8X = 1.69 - 0.09$

   $\quad 8X = 1.6$

   $\quad X = 1.6 \div 8$

   $\quad X = 0.2$

8. $8(0.2) + 0.09 = 1.69$

   $\quad 1.6 + 0.09 = 1.69$

   $\quad\quad 1.69 = 1.69$

9. $3.14(3.1 \text{ ft})^2 = 30.1754 \text{ ft}^2$

10. $3.14(6.2 \text{ ft}) = 19.468 \text{ ft}$

11. $4 \times 4 = 16$

12. $6 \times 6 = 36$

13. $5 \times 5 \times 5 = 125$

14. $4.2 \text{ m} + 5 \text{ m} + 4.2 \text{ m} + 5 \text{ m} = 18.4 \text{ m}$

15. $18 \text{ mm} + 9 \text{ mm} + 18 \text{ mm} + 9 \text{ mm} = 54 \text{ mm}$

16. $2.4 \text{ in} + 1.8 \text{ in} + 2.4 \text{ in} + 1.8 \text{ in} = 8.4 \text{ in}$

17. $4.2 \text{ m} \times 4 \text{ m} = 16.8 \text{ m}^2$

Lesson Practice 26A

1. done

2. done

3. done

4. $\dfrac{7}{50}$

5. $\dfrac{20}{300} = \dfrac{1}{15}$

6. $\dfrac{240}{300} = \dfrac{4}{5}$

7. $\dfrac{40}{300} = \dfrac{2}{15}$

8. $\dfrac{300-20}{300} = \dfrac{280}{300} = \dfrac{14}{15}$

9. $\dfrac{2}{1000} = \dfrac{1}{500}$

10. $\dfrac{10}{1000} = \dfrac{1}{100}$

11. $\dfrac{5}{500} = \dfrac{1}{100}$

12. $\dfrac{32}{192+32} = \dfrac{32}{224} = \dfrac{1}{7}$

Lesson Practice 26B

1. $\dfrac{25}{80} = \dfrac{5}{16}$

2. $\dfrac{20+15}{80} = \dfrac{35}{80} = \dfrac{7}{16}$

3. $\dfrac{20+25+15}{80} = \dfrac{60}{80} = \dfrac{3}{4}$

4. $\dfrac{20+25+13}{80} = \dfrac{58}{80} = \dfrac{29}{40}$

5. $\dfrac{10}{120} = \dfrac{1}{12}$

6. $\dfrac{100}{120} = \dfrac{5}{6}$

7. $\dfrac{10}{120} = \dfrac{1}{12}$

8. $\dfrac{100+10}{120} = \dfrac{110}{120} = \dfrac{11}{12}$

9. $\dfrac{4}{10} = \dfrac{2}{5}$

10. $\dfrac{7}{10}$

11. $\dfrac{4}{10} = \dfrac{2}{5}$

12. $\dfrac{3}{10}$

Lesson Practice 26C

1. $\dfrac{30}{250} = \dfrac{3}{25}$

2. $\dfrac{50+40+30+30}{250} = \dfrac{150}{250} = \dfrac{3}{5}$

3. $\dfrac{40+30}{250} = \dfrac{70}{250} = \dfrac{7}{25}$

4. $\dfrac{100+50+30}{250} = \dfrac{180}{250} = \dfrac{18}{25}$

5. $\dfrac{10}{3500} = \dfrac{1}{350}$

6. $\dfrac{500}{3500} = \dfrac{1}{7}$

7. $\dfrac{3500-500}{3500} = \dfrac{3000}{3500} = \dfrac{6}{7}$

8. $\dfrac{4}{10} = \dfrac{2}{5}$

9. $\dfrac{7}{10}$

10. $\dfrac{1}{10}$

11. $\dfrac{2}{10} = \dfrac{1}{5}$

12. $\dfrac{568 \div 2}{568} = \dfrac{284}{568} = \dfrac{1}{2}$

Systematic Review 26D

1. $\dfrac{60}{540} = \dfrac{1}{9}$

2. $\dfrac{450}{540} = \dfrac{5}{6}$

3. $\dfrac{450+60}{540} = \dfrac{510}{540} = \dfrac{17}{18}$

4. $\dfrac{5+9+9+9+13+15}{6} = 10$

mean = 10
median = 9
mode = 9

5. $\dfrac{1+1+3+3+3+3+5+6+6+9}{10} = 4$

mean = 4
median = 3
mode = 3

6. $\dfrac{34}{100} = \dfrac{17}{50}$

7. $\dfrac{178}{1000} = \dfrac{89}{500}$

8. $0.75 = 75\%$

9. $0.55\dfrac{5}{9} = 55\dfrac{5}{9}\%$

10. $5+4=9$; $9 \times 8 = 72$; $72-2 = 70$;
$70 \div 10 = 7$; $7+3 = 10$

11. $7-3=4$; $4 \times 6 = 24$; $24 \div 3 = 8$;
$8 \times 9 = 72$

12. $\dfrac{2}{5000} = \dfrac{1}{2500}$

13. $45\,lb \div 0.9lb\,/\,cust. = 50$ customers

14. $\$48.00 \times 0.20 = \$9.60$
$\$48.00 - \$9.60 = \$38.40$

Systematic Review 26E

1. $\dfrac{100}{500} = \dfrac{1}{5}$

2. $\dfrac{250+100}{500} = \dfrac{350}{500} = \dfrac{7}{10}$

3. $\dfrac{30+40+100}{500} = \dfrac{170}{500} = \dfrac{17}{50}$

4. $\dfrac{7+8+10+10+14+17}{6} = 11$

mean = 11
median = 10
mode = 10

5. $\dfrac{15+21+22+27+27}{5} = 22.4$

mean = 22.4
median = 22
mode = 27

6. $\dfrac{48}{100} = \dfrac{12}{25}$

7. $\dfrac{525}{1000} = \dfrac{21}{40}$

8. $0.83\dfrac{1}{3} = 83\dfrac{1}{3}\%$

9. $0.91\dfrac{2}{3} = 91\dfrac{2}{3}\%$

10. $36{,}320 \text{ g} \times \dfrac{1 \text{ kg}}{1{,}000 \text{ g}} = 36.32 \text{ kg}$

11. $\pi(10 \text{ ft})^2 \approx 314 \text{ ft}^2$

12. $25.02 \text{ gal} + 20.6 \text{ gal} + 0.25 \text{ gal} = 45.87 \text{ gal}$

13. $125.5 \text{ ft} \times 200 \text{ ft} = 25{,}100 \text{ ft}^2$

14. $9 + 4 = 13$; $13 + 3 = 16$; $16 + 4 = 20$;
$20 \div 5 = 4$; $4 - 0 = 4$

15. $10 \div 2 = 5$; $5 + 2 = 7$; $7 \times 7 = 49$;
$49 + 1 = 50$; $50 - 20 = 30$

Systematic Review 26F

1. $\dfrac{2}{10} = \dfrac{1}{5}$

2. $\dfrac{9}{10}$

3. $\dfrac{4}{10} = \dfrac{2}{5}$

4. $\dfrac{17 + 17 + 19 + 22 + 25}{5} = 20$

   mean $= 20$
   median $= 19$
   mode $= 17$

5. $\dfrac{1 + 1 + 1 + 7 + 7 + 8 + 9 + 14}{8} = 6$

   mean $= 6$
   median $= 7$
   mode $= 1$

6. $\dfrac{65}{100} = \dfrac{13}{20}$

7. $\dfrac{125}{1000} = \dfrac{1}{8}$

8. $0.42\dfrac{6}{7} = 42\dfrac{6}{7}\%$

9. $0.90 = 90\%$

10. $200 \times 0.87 = 174$

11. $1 \text{ km} \times \dfrac{1{,}000 \text{ m}}{1 \text{ km}} \times \dfrac{1 \text{ hm}}{100 \text{ m}} = 10 \text{ hm}$;

    10 runners

12. $3.14(12 \text{ ft}) = 37.68 \text{ ft}$; 38 lengths

13. meter

14. $\dfrac{1}{2} \times 8 = 4$; $4 + 3 = 7$;
$7 \times 7 = 49$; $49 + 1 = 50$

15. $8 + 9 = 17$; $17 - 7 = 10$;
$10 \div 5 = 2$; $2 \times 4 = 8$

Lesson Practice 27A

1. ray
2. line segment
3. geometry
4. point
5. line
6. d: line
7. c: ray
8. b: line segment
9. e: point
10. a: infinity
11. A
12. $\overleftrightarrow{ST}$ or $\overleftrightarrow{TS}$

Lesson Practice 27B

1. ray
2. line segment
3. endpoint
4. points
5. point
6. d
7. a
8. c
9. b
10. $\overline{LM}$ or $\overline{ML}$
11. $\overrightarrow{QR}$
12. line

Lesson Practice 27C

1. point
2. line
3. ray
4. line segment
5. ray
6. e
7. d
8. a
9. c
10. b
11. $\overleftrightarrow{DE}$ or $\overleftrightarrow{ED}$
12. $\overline{YZ}$ or $\overline{ZY}$

Systematic Review 27D

1. length; width
2. length; width
3. one
4. two
5. $\overrightarrow{EF}$
6. $3+5+5+5+6+7+7+9+16=63$
   mean $=63 \div 9 = 7$
7. median $=6$
8. mode $=5$
9. $\dfrac{15}{100} = \dfrac{3}{20}$
10. $\dfrac{375}{1000} = \dfrac{3}{8}$
11. $0.6 + 0.045 = 0.645$
12. $113 + 0.19 = 113.19$
13. $2.9 + 0.11 = 3.01$
14. 76
15. 55
16. $\dfrac{1}{12}$
17. $\dfrac{1}{8}$
18. $10 \div 2 = 5; \ 5 \times 4 = 20;$
    $20 \div 2 = 10; \ 10 + 7 = 17$

Systematic Review 27E

1. line; ray
2. point
3. line segment
4. endpoint
5. $\overleftrightarrow{GK}$ or $\overleftrightarrow{KG}$
6. $1+1+1+7+7+8+9+14=48$
   mean $=48 \div 8 = 6$
7. median $=7$
8. mode $=1$
9. $\dfrac{65}{100} = \dfrac{13}{20}$
10. $\dfrac{1}{1000}$
11. $11.4 - 0.6 = 10.8$
12. $35 - 0.42 = 34.58$
13. $5.15 - 0.9 = 4.25$
14. 890
15. north

16. $\dfrac{15}{150} = \dfrac{1}{10}$
17. $3.14 \times 12{,}732 \text{ km} \approx 39{,}978 \text{ km}$
18. $4 \times 2 = 8; \ 8 + 1 = 9;$
    $9 - 7 = 2; \ 2 \div 2 = 1$

Systematic Review 27F

1. infinite
2. point
3. endpoints
4. endpoint or origin
5. $\overline{MN}$ or $\overline{NM}$
6. $13+13+13+13+14+18=84$
   mean $=84 \div 6 = 14$
7. median $=13$
8. mode $=13$
9. $\dfrac{82}{100} = \dfrac{41}{50}$
10. $\dfrac{5}{1000} = \dfrac{1}{200}$
11. $0.5 \times 0.7 = 0.35$
12. $0.02 \times 0.4 = 0.008$
13. $1.31 \times 0.28 = 0.3668$
14. because it goes the whole way around the city
15. $\dfrac{1}{7}$
16. $4.5 \text{ m} \times 1.9 \text{ m} \times 6 \text{ m} = 51.3 \text{ m}^3$
17. $7 \text{ ft} \times 1.12 = 7.84 \text{ ft}$
18. $\dfrac{1}{3} \times 12 = 4; \ 4 + 5 = 9;$
    $9 + 1 = 10; \ 10 \times 4 = 40$

Lesson Practice 28A

1. length; width
2. two
3. plane
4. congruent
5. equal
6. similar
7. b
8. a
9. d

10. c
11. g
12. e
13. h
14. f

Lesson Practice 28B

1. equal
2. similar
3. congruent
4. plane
5. point
6. geometry
7. d
8. c
9. b
10. a
11. h
12. g
13. e
14. f

Lesson Practice 28C

1. point
2. line, line segment, ray
3. plane
4. equal
5. congruent
6. similar
7. c
8. d
9. a
10. b
11. ▱
12. =
13. ~
14. ≅

Systematic Review 28D

1. congruent
2. length; width
3. similar
4. length; width
5. one
6. length; width
7. two

8. $0.7T + 1.4 = 4.9$
   $0.7T = 4.9 - 1.4$
   $0.7T = 3.5$
   $T = 3.5 \div 0.7$
   $T = 5$

9. $0.7(5) + 1.4 = 4.9$
   $3.5 + 1.4 = 4.9$
   $4.9 = 4.9$

10. $33\frac{1}{3}\%$
11. 50%
12. 75%
13. $0.15 \div 0.9 = 0.17$ 20 ✗ ⟨.17⟩
14. $1.28 \div 6.2 = 0.21$
15. $0.105 \div 0.5 = 0.21$
16. $3 + 5 + 6 + 6 + 10 = 30$
    $30 \div 5 = 6$
17. 90
18. $6 \div 3 = 2$; $2 \times 4 = 8$;
    $8 - 3 = 5$; $5 + 6 = 11$

Systematic Review 28E

1. h
2. e
3. b
4. f
5. d
6. c
7. a
8. i
9. g
10. $66\frac{2}{3}\%$
11. ⟨25%⟩ 3⟩⟨3 ⟨25%⟩
12. 20%

13.
$$.9133 = 0.91\overline{3}$$
$$3\overline{)2.7400}$$
$$\underline{27000}$$
$$400$$
$$\underline{300}$$
$$100$$
$$\underline{900}$$
$$10$$
$$\underline{9}$$
$$1$$

14.
$$90.66 = 90.\overline{6}$$
$$9.\overline{)816.00}$$
$$\underline{81000}$$
$$600$$
$$\underline{540}$$
$$60$$
$$\underline{54}$$
$$6$$

15.
$$4.145 = 4.1\overline{45}$$
$$11.\overline{)45.600}$$
$$\underline{44000}$$
$$1600$$
$$\underline{1100}$$
$$500$$
$$\underline{440}$$
$$60$$
$$\underline{55}$$
$$5$$

16. 251

17. 176

18. $9 \div 3 = 3$; $3 + 2 = 5$;
$5 \times 7 = 35$; $35 - 5 = 30$

Systematic Review 28F

1. ∞
2. $-$
3. $\rightarrow$
4. $\leftrightarrow$
5. $\bullet$
6. $=$
7. $\sim$
8. $\cong$
9. $\square$

10. 40%

11. $33\dfrac{1}{3}$%

12. 80%

13. $0.73\dfrac{13}{19}$

14. $0.85\dfrac{5}{7}$

15. $0.55\dfrac{5}{9}$

16. 12 in

17. north; 435

18. $\dfrac{1}{4} \times 12 = 3$; $3 + 7 = 10$;
$10 \div 2 = 5$; $5 \times 5 = 25$

Lesson Practice 29A

1. angles
2. rays
3. 90
4. 360
5. vertex
6. circle or turn
7. Greek
8. done
9. $\angle$QXT; $\angle$TXQ
10. 90°
11. four

Lesson Practice 29B

1. angle
2. right
3. endpoint
4. circle or turn
5. right angle
6. lines
7. vertex
8. $\angle$ZYX; $\angle$XYZ
9. $\angle$LMN; $\angle$NML
10. right
11. 90°

Lesson Practice 29C

1. 90
2. 360
3. vertex
4. Greek
5. $\dfrac{1}{4}$
6. 4
7. rays
8. $\angle CBA$; $\angle ABC$
9. $\angle YRD$; $\angle DRY$
10. 90°
11. 90°

Systematic Review 29D

1. circle
2. 360°
3. 90°
4. similar
5. length; width
6. two
7. congruent
8. $0.85M = 0.425$
   $M = 0.425 \div 0.85$
   $M = 0.5$
9. $0.85(0.5) = 0.425$
   $0.425 = 0.425$
10. $\dfrac{95}{100} = \dfrac{19}{20}$
11. $\dfrac{7}{10}$
12. $\dfrac{685}{1000} = \dfrac{137}{200}$
13. $32 \div 8 = 4$
    $4 \times 3 = 12$
14. $12 \div 6 = 2$
    $2 \times 1 = 2$
15. $72 \div 9 = 8$
    $8 \times 5 = 40$
16. $\dfrac{950-2}{950} = \dfrac{948}{950} = \dfrac{474}{475}$
17. $8.5 \text{ lb} + 9.5 \text{ lb} + 10 \text{ lb} + 10 \text{ lb} + 12 \text{ lb} = 50 \text{ lb}$
    $50 \text{ lb} \div 5 = 10 \text{ lb}$

18. $\dfrac{1}{5} \times 15 = 3$; $3 + 8 = 11$;
    $11 - 2 = 9$; $9 \times 9 = 81$

Systematic Review 29E

1. b
2. c
3. f
4. e
5. a
6. g
7. d
8. $0.09Q = 0.315$
   $Q = 0.315 \div 0.09 = 3.5$
9. $0.09(3.5) = 0.315$
   $0.315 = 0.315$
10. $\dfrac{24}{100} = \dfrac{6}{25}$
11. $\dfrac{9}{10}$
12. $\dfrac{278}{1000} = \dfrac{139}{500}$
13. $6\dfrac{1}{6} + 8\dfrac{2}{5} = 6\dfrac{5}{30} + 8\dfrac{12}{30} = 14\dfrac{17}{30}$
14. $3 - 2\dfrac{1}{3} = 2\dfrac{3}{3} - 2\dfrac{1}{3} = \dfrac{2}{3}$
15. $7\dfrac{2}{9} + 9\dfrac{4}{5} = 7\dfrac{10}{45} + 9\dfrac{36}{45} =$
    $16\dfrac{46}{45} = 17\dfrac{1}{45}$
16. $3,405 \text{ g} \times \dfrac{1 \text{ kg}}{1,000 \text{ g}} = 3.405 \text{ kg}$
17. south
18. $20 \div 10 = 2$; $2 \times 6 = 12$;
    $12 + 3 = 15$; $15 - 8 = 7$

Systematic Review 29F

1. right
2. $\dfrac{1}{4}$
3. length; width
4. equal
5. line segment
6. rays
7. $\angle BCA$; $\angle ACB$

8. $3.3X + 0.79 = 4.42$
   $3.3X = 4.42 - 0.79$
   $3.3X = 3.63$
   $X = 3.63 \div 3.3$
   $X = 1.1$

9. $3.3(1.1) + 0.79 = 4.42$
   $3.63 + 0.79 = 4.42$
   $4.42 = 4.42$

10. $\dfrac{75}{100} = \dfrac{3}{4}$

11. $\dfrac{9}{100}$

12. $\dfrac{138}{1000} = \dfrac{69}{500}$

13. $\dfrac{\cancel{4}}{9} \times \dfrac{5}{\cancel{8}_2} = \dfrac{5}{18}$

14. $\dfrac{\cancel{2}}{3} \times \dfrac{\cancel{9}^3}{\cancel{10}_5} = \dfrac{3}{5}$

15. $\dfrac{3}{\cancel{24}} \times \dfrac{\cancel{2}}{5} = \dfrac{3}{10}$

16. $\$175.62 - \$39.59 = \$136.03$

17. $2.75 \text{ gal/min} \times 12.25 \text{ min} = 33.6875 \text{ gal}$

18. $8 \div 4 = 2; \ 2 \times 5 = 10;$
    $10 - 6 = 4; \ 4 + 9 = 13$

Lesson Practice 30A

1. acute
2. obtuse
3. straight
4. b
5. d
6. a
7. c
8. right
9. obtuse
10. acute
11. straight
12. straight

Lesson Practice 30B

1. straight
2. obtuse
3. acute

4. c
5. b
6. d
7. a
8. straight
9. obtuse
10. right
11. acute
12. $360° \div 6 = 60°$; acute

Lesson Practice 30C

1. 90°; 180°
2. 0°; 90°
3. 180°
4. 90°
5. b
6. c
7. a
8.
9.
10.
11.
12. 180°

Systematic Review 30D

1. acute
2. length; width
3. 90°
4. similar
5. congruent
6. two
7. vertex
8. $0.07X + 0.52 = 0.527$
   $0.07X = 0.527 - 0.52$
   $0.07X = 0.007$
   $X = 0.1$
9. $0.07(0.1) + 0.52 = 0.527$
   $0.007 + 0.52 = 0.527$
   $0.527 = 0.527$

10. $5 \div 6 = 0.83$

11. $8 \div 13 = 0.62$

12. $4 \div 5 = 0.8$

13. $\dfrac{1}{3} \div \dfrac{4}{7} = \dfrac{1}{3} \times \dfrac{7}{4} = \dfrac{7}{12}$

14. $\dfrac{3}{10} \div \dfrac{3}{5} = \dfrac{\cancel{3}}{\cancel{10}_2} \times \dfrac{\cancel{5}}{\cancel{3}} = \dfrac{1}{2}$

15. $\dfrac{7}{8} \div \dfrac{1}{2} = \dfrac{7}{\cancel{8}_4} \times \dfrac{\cancel{2}}{1} = \dfrac{7}{4} = 1\dfrac{3}{4}$

16. $3.14(26 \text{ in}) 81.64 \text{ in}$

17. $\dfrac{200}{200+400} = \dfrac{200}{600} = \dfrac{1}{3}$

18. $30 \div 10 = 3$

$3 \times 4 = 12$

$12 + 2 = 14$

$14 - 7 = 7$

Systematic Review 30E

1. b

2. e

3. f

4. c

5. a

6. d

7. g

8.

9.

10. $\dfrac{725}{1000} = \dfrac{29}{40}$

11. $\dfrac{3}{10}$

12. $\dfrac{95}{100} = \dfrac{19}{20}$

13. $65 \text{ l} \times \dfrac{1 \text{ kl}}{1,000 \text{ l}} = 0.065 \text{ kl}$

14. $21 \text{ mm} \times \dfrac{1 \text{ m}}{1,000 \text{ mm}} \times \dfrac{100 \text{ cm}}{1 \text{ m}} = 2.1 \text{ cm}$

15. $0.04 \text{ g} \times \dfrac{100 \text{ cg}}{1 \text{ g}} = 4 \text{ cg}$

16. round: $3.14(6 \text{ in})^2 = 113.04 \text{ in}^2$

square: $10.5 \text{ in} \times 10.5 \text{ in} = 110.25 \text{ in}^2$

$113.04 \text{ in}^2 > 110.25 \text{ in}^2$

round pizza is bigger

17. $\$5.95 \times 1.03 \approx \$6.13$

$\$10.00 - \$6.13 = \$3.87$

18. $10 \times 4 = 40$

$40 \div 5 = 8$

$8 \div 4 = 2$

$2 \div 2 = 1$

Systematic Review 30F

1. acute

2. obtuse

3. length; width

4. equal

5. $\dfrac{1}{4}$

6. congruent

7. $\angle XTQ; \angle QTX$

8. $3+5+5+5+6+7+7+9+16 = 63$

mean $= 63 \div 9 = 7$

median $= 6$

mode $= 5$

9. $9+13+27+27+30+32+37 = 175$

mean $= 175 \div 7 = 25$

median $= 27$

mode $= 27$

10. 350%

11. 1%

12. 45%

13. $0.35 \times 200 = 70$

14. $0.02 \times 16 = 0.32$

15. $2.50 \times 150 = 375$

16. $\dfrac{5}{695} = \dfrac{1}{139}$

17. $6^2 = 6 \times 6 = 36$

18. $8 + 4 = 12$

$12 - 6 = 6$

$6 + 3 = 9$

$9 \times 2 = 18$

Application & Enrichment Solutions

Application & Enrichment 1G

1. Sum
2. Factors
3. Product
4. Quotient
5. G, B
6. F, optional: C, E
7. A, optional: E
8. H, optional: E
9. E, optional: A
10. B, optional: E
11. C, optional: E
12. D, optional: E
13. B, optional: A, E
14. 3X
15. 3G
16. 4Y
17. 11A

Application & Enrichment 2G

1. Each expression has a value of 9
2. Always True
3. Always True
4. Sometimes False
5. Always True
6. Sometimes False
7. Done
8. $20 \times 6 = 120$
   $4 \times 30 = 120$
9. $6 + 7 = 13$
   $9 + 4 = 13$
10. $2 \div 2 = 1$
    $12 \div 3 = 4$
11. $5 - 3 = 2$
    $10 - 2 = 8$

Application & Enrichment 3G

1. 4
2. 5
3. 8
4. 9
5. 5

6. 1, 2, 3, 4, 6, 8, 12, 24
7. 1, 3, 9
8. 1, 3, 5, 9, 15, 45
9. 1, 2, 4, 8, 16
10. $6(1) + 6(2)$
11. $5(5) + 5(6)$
12. $3(3) + 3(7)$
13. $6(1 + 2) = 6(3) = 18$
    $6 + 12 = 18$
14. $5(5 + 6) = 5(11) = 55$
    $25 + 30 = 55$
15. $3(3 + 7) = 3(10) = 30$
    $9 + 21 = 30$

Application & Enrichment 4G

1. $3(7) = 21$
   $18 + 3 = 21$ The answers agree.
2. $4(2 + 5) = 4(7) = 28$
   $4(2) + 4(5) = 8 + 20 = 28$
3. $2(10 + 6) = 2(16) = 32$
   $2(10) + 2(6) = 20 + 12 = 32$
4. $6(2 + 3) = 6(5) = 30$
   $6(2) + 6(3) = 12 + 18 = 30$
5. $7(8 + 4) = 7(12) = 84$
   $7(8) + 7(4) = 56 + 28 = 84$
6. $4(2) + 4(B) = 8 + 4B$
7. $8(A) + 8(4) = 8A + 32$
8. $2(X) + 2(Z) = 2X + 2Z$
9. $7(A) + 7(2Y) = 7A + 14Y$
10. $2(3) + 2(5B) = 2(3 + 5B)$
11. $5(3) + 5(7X) = 5(3 + 7X)$
12. $9(1) + 9(8Y) = 9(1 + 8Y)$

Application & Enrichment 5G

1. Answers will vary. Both sides of final equation should match. Example:
   $1 + 1 = 2(1); 2 = 2$
   $5 + 5 = 2(5); 10 = 10$
2. Answers will vary. Both sides of final equation should match.
3. Answers will vary. Both sides of final equation should match.

4. 3(2) + 4 = 4 + 3(2)
6 + 4 = 4 + 6
10 = 10; true

5. 5(8 − 6) = 5(8) − 30
5(2) = 40 − 30
10 = 10; true

6. 4(3) =2(3) + 2
12 = 6 + 2
12 ≠ 8; false

7. 5(2) ÷ 5 = 5 ÷ 5(2)
10 ÷ 5 = 5 ÷ 10
2 ≠ 1/2; false

8. Done

9. 3, 6, 9, 12, 15 . . .
4, 8, 12, 16 . . .
The LCM of 3 and 4 is 12.

10. 6, 12, 18, 24, 30 . . .
10, 20, 30, 40 . . .
The LCM of 6 and 10 is 30.

11. 4, 8, 12, 16 . . .
12, 24, 36, 48 . . .
The LCM of 4 and 12 is 12.

12. 9, 18, 27, 36, 45 . . .
12, 24, 36, 48 . . .
The LCM of 9 and 12 is 36.

Application & Enrichment 6G

1. 10 years
2. 24 years
3. 14 years
4. 10, 15, 24 years
5-8. On graph, 1 has 1 dot, 2 has 2, 3 has 3, 7 has 1 dot
5. 3
6. 1, 7
7. 1 + (2 × 2) + (3 × 3) + 7 = 1 + 4 + 9 + 7 = 21 pets
8. 21 ÷ 7 = 3 pets average

Application & Enrichment 7G

1. 3 kg
2. 5 − 3 = 2 kg
3. 8 kg

4. Birth to three months
5.

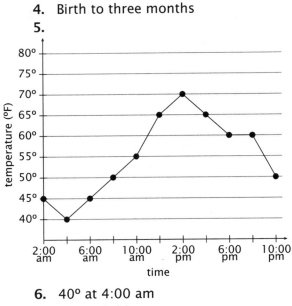

6. 40° at 4:00 am
7. 70° at 2:00 pm

Application & Enrichment 8G

1. Freddie, 4 flies
2. Billy, 14 flies
3. 4 + 8 + 6 + 14 = 32 flies
4. 32 ÷ 4 = 8 flies
5. no
6.

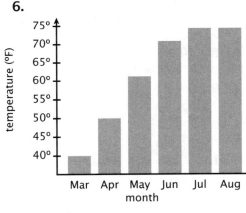

7. March
8. July, August
9. 74° − 40° = 34°

Application & Enrichment 9G

1. 31–40 years

2. 61–70 years

3. Yes, because many of the employees are around the age at which people are most likely to have young families (mid 20's to late 30's). Answers may vary.

4. No, because the fact that employees are the right age to have young families does not necessarily mean they are married or have children. Answers may vary.

5. 60°–65°: 6 times
65°–70°: 3 times
70°–75°: 9 times
75°–80°: 1 time
80°–85°: 1 time

6.

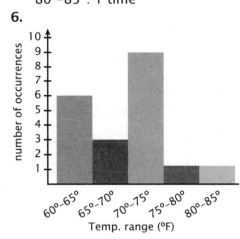

Application & Enrichment 10G

1. $3.29
2. $16.12
3. $5.70
4. $0.35
5. 25.4 inches
6. 6.92 m
7. 125 mi
8. 0.40 cm
9. 1,568.21 miles
10. 1,600 miles

Application & Enrichment 11G

1.

trunks	legs
1	4
2	8
3	12
4	16

2. Counting by one/adding one
3. Skip counting by 4/multiply by 4
4. 4

dishes	minutes
3	1
6	2
9	3
12	4

5. 8 minutes. Divide by 3 or multiply by 1/3.
6. 8 dishes ÷ 4 min = 2 dishes/min

Application & Enrichment 12G

1. 1/10 × 300 = 30 votes
2. 16 red for every 20 yellow

red	4	8	12	16
yellow	5	10	15	20

3. 8/4 = 8 ÷ 4 = 2 minutes per pan
4. 4 × 15 = 60
2 × 4 = 8 incorrect

Application & Enrichment 13G

1.

gallons of green	2	4	6	8	10
total gallons	3	6	9	12	15

2. 8 gallons
3. 12 − 8 = 4 gallons
4.

cranberry	3	6	9	12	15
mixture	5	10	15	20	25

5. 12 cups
6. 20 − 12 = 8 cups

Application & Enrichment 14G

1.

miles	hours	mph
200	4	50
300	6	50
150	3	50
400	8	50

2.

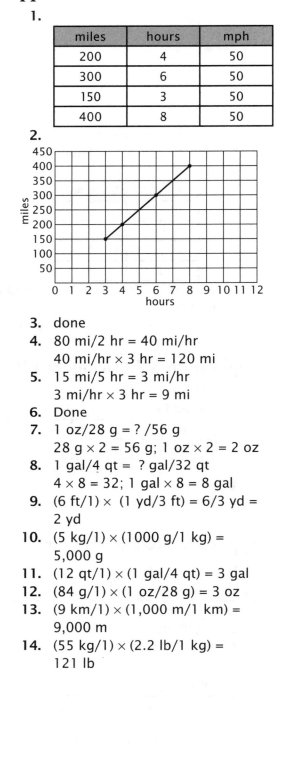

3. done

4. 80 mi/2 hr = 40 mi/hr
   40 mi/hr × 3 hr = 120 mi

5. 15 mi/5 hr = 3 mi/hr
   3 mi/hr × 3 hr = 9 mi

6. Done

7. 1 oz/28 g = ? /56 g
   28 g × 2 = 56 g; 1 oz × 2 = 2 oz

8. 1 gal/4 qt = ? gal/32 qt
   4 × 8 = 32; 1 gal × 8 = 8 gal

9. (6 ft/1) × (1 yd/3 ft) = 6/3 yd = 2 yd

10. (5 kg/1) × (1000 g/1 kg) = 5,000 g

11. (12 qt/1) × (1 gal/4 qt) = 3 gal

12. (84 g/1) × (1 oz/28 g) = 3 oz

13. (9 km/1) × (1,000 m/1 km) = 9,000 m

14. (55 kg/1) × (2.2 lb/1 kg) = 121 lb

Application & Enrichment 15G

1.

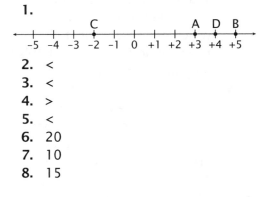

2. <

3. <

4. >

5. <

6. 20

7. 10

8. 15

Application & Enrichment 16G

1. $55

2. −32 > −120; Justin is camping at the higher location.

3. Jerry; the absolute value (distance from zero) of −120 is greater than the absolute value of −32

4. 31° > −13°

5. |31| > |−13| or |−13| < |31|

6.

7.

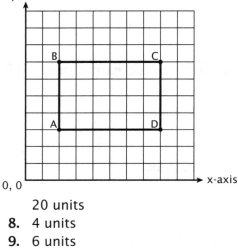

20 units

8. 4 units

9. 6 units

10. 6 + 6 + 4 + 4 = 20 units
Adding the lengths of the sides
yields the same answer as
counting.

Application & Enrichment 17G

1.

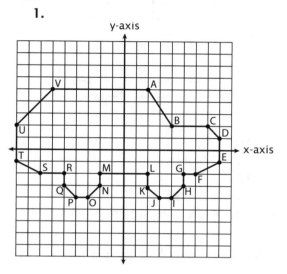

5.

feet	minutes
3	1
6	2
9	3
12	4
15	5

6. 2
7. 3
8. 4
9. skip counting by 3
10.

Application & Enrichment 18G

1.

wheat flour	rolled oats
5	2
10	4
15	6
20	8

2. 5 cups
3. 2 cups
4.

Application & Enrichment 19G

1. 0.8A = 160
A = 160 ÷ 0.8; A = 200 yd²
2. 0.7D = 49
D = 49 ÷ 0.7; D = 70 in
3. 0.25(M) = $15
M = $15 ÷ 0.25 = $60
4. 0.75P = $261
P = $261 ÷ 0.75 = $348
5. 0.3T = 3
T = 3 ÷ 0.3 = 10 hours
6. false, false, true
7. true, true, false
8. 0.5(0.4) = 0.2; <
9. 0.5(0.4) = 2; <
10. 0.5(40) = 20; >

Application & Enrichment 20G

1. $2\pi r \approx 2(3.14)(2) \approx 12.56$ in
   $\pi(2^2) \approx 12.56$ in^2
2. $2\pi r \approx 2(3.14)(4) \approx 25.12$ in
   $\pi(4^2) \approx 3.14(16) \approx 50.24$ in
3. 2
4. 4
5. $2\pi r \approx 2(3.14)(3) \approx 18.84$ in
   $\pi(3^2) \approx 3.14(9) \approx 28.26$ in^2
6. $2\pi r \approx 2(3.14)(6) \approx 37.68$ in
   $\pi(6^2) \approx 3.14(36) \approx 113.04$ in^2
7. 2
8. 4

Application & Enrichment 21G

1. (4, 0)
2. (4, 5)
3. (0, 5)
4. (0, 0)
5.

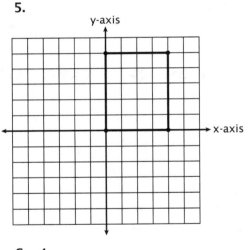

6. 4
7. 5
8. $4 \times 5 = 20$

9.

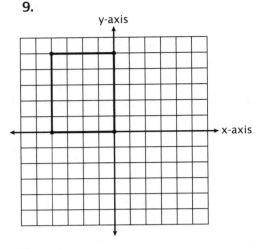

10. –4
11. 4 blocks
12. 5
13. 5 blocks
14. $4 \times 5 = 20$ square blocks

Application & Enrichment 22G

1. $5C = 20$
2. $C = 4$ cars
3. $\$3.50(D) = M$
   $\$3.50(5) = \$17.50$
4. $\$3.50(12) = \$42$
5. $\$3.50(D) = \$35$
   $D = 10$ days and sandwiches
6.

radius	1 cm	2 cm	4 cm	8 cm
$C = 2(3.14)r$	6.28 cm	12.56 cm	25.12 cm	50.24 cm

7.

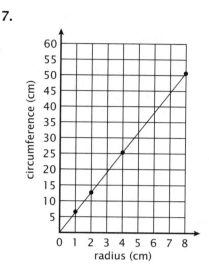

Application & Enrichment 23G

1. 3 ft × 4 ft × 5 ft = 60 ft³
2. bottom = 15 ft²
   top = 15 ft²
   front = 20 ft²
   back = 20 ft²
   side 1 = 12 ft²
   side 2 = 12 ft²
3. Surface area = 94 ft²
4. 96 ft² + 120 ft² + 96 ft² + 120 ft²
   + 80 ft² + 80 ft² = 592 ft²
5. 600 ft² ×(1 can/100 ft²) = 6 cans
6. Triangles:
   (1/2)(6 in × 4 in) = 12 in²
   Square: 6 in × 6 in = 36 in²
   (12 in² × 4) + 36 in² = 84 in²

Application & Enrichment 24G

1. 6
2. 12
3. yes
4. 6 + 6 + 6 + 6 + 4 + 4 = 32
   squares
5. 1.5 in × 1.5 in = 2.25 in²

6.

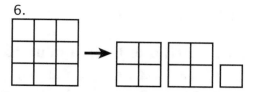

7. (1.5 in)(1.5 in)(1 in) = 2.25 in³
   Since one unit block is about half
   an inch on a side, it takes about
   eight to make a cubic inch. You
   should have used 18 blocks to
   build the shape, since 18 ÷ 8 =
   2.25.
8. (1.5 in)(1.5 in)(1.5 in) = 3.375 in³
   27 blocks = 3 groups of eight
   blocks (3 in³) with 3 left over
   (0.375 in³)
9. T < 100° C
10.

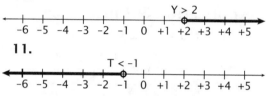

11.

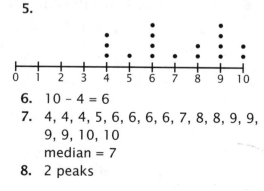

Application & Enrichment 25G

1. lowest = 2
   highest = 10
   spread = 10 − 2 = 8
2. 2, 3, 4, 5, 5, 6, 6, 7, 7, 7, 8, 8, 8,
   8, 9, 9, 10
   median = 7
3. mode = 8
4. asymmetrical
5.
6. 10 − 4 = 6
7. 4, 4, 4, 5, 6, 6, 6, 6, 7, 8, 8, 9, 9,
   9, 9, 10, 10
   median = 7
8. 2 peaks

9. $120 \div 17 = 7.1$

Application & Enrichment 26G

1. $4 + 2 + 4 + 6 + 2 + 1 = 19$ people
2. $2 + 4 + 2 + 4 + 1 + 1 = 14$ people
3. $19 + 14 = 33$ people
4. no
5. B
6. B
7. Answers will vary. For example, "Do you support spending town funds for the improvement of Main Street?"
8. inches
9. feet
10. Answers will vary.

Application & Enrichment 27G

1. $(1 + 3 + 3 + 5 + 6 + 9) \div 6 = 27 \div 6 = 4.5$ average per family

2.

J	L	B	R	T	M
1	3	3	5	6	9

3.

J	L	B	R	T	M
1	3	3	5	6	9
3.5	1.5	1.5	0.5	1.5	4.5

4. $(3.5 + 1.5 + 1.5 + 0.5 + 1.5 + 4.5) \div 6 = 13 \div 6 = 2\ 1/6 \approx 2.17$

Application & Enrichment 28G

1. 2, 3, 3, 3, 3, 4, 4, 4, 5, 5, 6, 7, 14; median = 4
2. 2, 3, 3, 3, 3, 4
    1st quartile = $(3 + 3) \div 2 = 3$
3. 4, 5, 5, 6, 7, 14
    3rd quartile = $(5 + 6) \div 2 = 5.5$
4. IQR = $5.5 - 3 = 2.5$
5. Range = $14 - 2 = 12$ much more than twice the size of the IQR

even though it is only twice as many data.

6. Mean = $(2 + 3 + 3 + 3 + 3 + 4 + 4 + 4 + 5 + 5 + 6 + 7 + 14) \div 13 = 4.8$, which is 0.8 longer than the median

Application & Enrichment 29G

1.

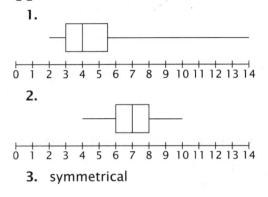

2.

3. symmetrical

Application & Enrichment 30G

The student should review the main ideas and use them when observing and recording real-life data.

Test Solutions

Lesson Test 1

1. $1 \times 1 \times 1 \times 1 = 1$
2. $10 \times 10 = 100$
3. $13 \times 13 = 169$
4. $3 \times 3 \times 3 = 27$
5. $12 \times 12 = 144$
6. $7 \times 7 = 49$
7. $5^2 = 25$
8. $4^3 = 64$
9. $2^4 = 16$
10. $6 \times 6 \times 6 \times 6 = 6^4$
11. $15 \times 15 = 15^2$
12. $9 \times 9 \times 9 = 9^3$
13. $8 \times 8 \times 8 = 8^3$ or $512 = 8^3$
14. $512 = 8^3$ or $8 \times 8 \times 8 = 8^3$
15. $96 \div 4 = 24$
    $24 \times 1 = 24$
16. $88 \div 8 = 11$
    $11 \times 3 = 33$
17. $100 \div 10 = 10$
    $10 \times 7 = 70$
18. $60 \text{ min} \div 4 = 15 \text{ min}$
    $15 \text{ min} \times 3 = 45 \text{ min}$
19. $24 \div 8 = 3$
    $3 \times 1 = 3$ apples
20. $20 \div 10 = 2$
    $2 \times 9 = 18$ people

Lesson Test 2

1. $10 \times 10 \times 10 \times 10 = 10,000$
2. $10 \times 10 = 100$
3. $10 \times 10 \times 10 = 1,000$
4. $10 = 10^1$
5. $1,000,000 = 10^6$
6. $1 = 10^0$
7. $1 \times 100 + 2 \times 10 + 5 \times 1$
8. $3 \times 1,000 + 9 \times 100 + 3 \times 10 + 2 \times 1$
9. $9 \times 10^3 + 2 \times 10^1 + 8 \times 10^0$
10. $1 \times 10^4 + 6 \times 10^3 + 9 \times 10^1$
11. $83,200$

12. $2,680$
13. $8 \times 8 = 64$
14. $2 \times 2 \times 2 = 8$
15. $5^3 = 125$
16. $\dfrac{1}{8} = \dfrac{2}{16} = \dfrac{3}{24} = \dfrac{4}{32}$
17. $\dfrac{4}{5} = \dfrac{8}{10} = \dfrac{12}{15} = \dfrac{16}{20}$
18. $24 \text{ hr} \div 6 = 4 \text{ hr}$
    $4 \text{ hr} \times 3 = 12 \text{ hr}$
19. $24 \text{ hr} - 12 \text{ hr} = 12 \text{ hr}$
20. $1/2$ of $60 \text{ min} = 30 \text{ min}$
    $1/3$ of $60 \text{ min} = 20 \text{ min}$
    $1/4$ of $60 \text{ min} = 15 \text{ min}$
    $1/4$ of $60 \text{ min} = 15 \text{ min}$
    $1/2$ of $60 \text{ min} = 30 \text{ min}$
    $30 + 20 + 15 + 15 + 30 = 110 \text{ min}$

Lesson Test 3

1. $3 \times 10^1 + 4 \times 10^0 + 6 \times \dfrac{1}{10^1}$
2. $1 \times \dfrac{1}{10^1} + 2 \times \dfrac{1}{10^2} + 4 \times \dfrac{1}{10^3}$
3. $5,417.04$
4. $1,000.875$
5. dimes; cents;
   $3.00 + 0.0 + 0.05 = 3.05$
6. dollars; dimes; cent;
   $6.00 + 0.4 + 0.01 = 6.41$
7. $10 \times 10 \times 10 = 1,000$
8. $1 \times 1 \times 1 \times 1 \times 1 \times 1 = 1$
9. $6 \times 6 = 36$
10. $11 \times 11 = 121$
11. $\dfrac{2}{9} = \dfrac{4}{18} = \dfrac{6}{27} = \dfrac{8}{36}$
12. $\dfrac{1}{7} = \dfrac{2}{14} = \dfrac{3}{21} = \dfrac{4}{28}$
13. $\dfrac{3}{12} = \dfrac{1}{4}$
14. $\dfrac{2}{50} = \dfrac{1}{25}$
15. $\dfrac{8}{64} = \dfrac{1}{8}$

16. $\dfrac{15}{90} = \dfrac{1}{6}$

17. $4.63

18. 100 cents ÷ 10 = 10 cents

10 cents × 5 = 50 cents

19. $\dfrac{50}{100} = \dfrac{1}{2}$

20. $5 \times 5 = 5^2$ blocks

Lesson Test 4

1.
$$
\begin{array}{r}
{}^{1} \\
6.7 \\
+\ 5.4 \\
\hline
12.1
\end{array}
$$

2.
$$
\begin{array}{r}
2.0 \\
+\ .2 \\
\hline
2.2
\end{array}
$$

3.
$$
\begin{array}{r}
6.24 \\
+\ 8.40 \\
\hline
14.64
\end{array}
$$

4.
$$
\begin{array}{r}
{}^{1} \\
5.28 \\
+2.05 \\
\hline
7.33
\end{array}
$$

5. $1 \times 1 \times 1 = 1$

6. $10 \times 10 = 100$

7. $6 \times 6 \times 6 = 216$

8. $7 \times 7 = 49$

9. 8,400.2

10. 269.005

11. $\dfrac{4}{5} = \dfrac{8}{10} = \dfrac{12}{15} = \dfrac{16}{20}$

12. $\dfrac{5}{9} = \dfrac{10}{18} = \dfrac{15}{27} = \dfrac{20}{36}$

13. $\dfrac{1}{6} + \dfrac{2}{9} = \dfrac{9}{54} + \dfrac{12}{54} = \dfrac{21}{54} = \dfrac{7}{18}$

14. $\dfrac{1}{5} + \dfrac{7}{10} = \dfrac{10}{50} + \dfrac{35}{50} = \dfrac{45}{50} = \dfrac{9}{10}$

15. $\dfrac{1}{3} + \dfrac{3}{8} = \dfrac{8}{24} + \dfrac{9}{24} = \dfrac{17}{24}$

16. $4.75 + $6.29 = $11.04

$11.04 > $11.00; yes

17. 1.2 + .75 + 1.15 = 3.1 carats

18. $\dfrac{4}{10} = \dfrac{2}{5}$

19. $\dfrac{5}{8} + \dfrac{1}{6} = \dfrac{30}{48} + \dfrac{8}{48} = \dfrac{38}{48} = \dfrac{19}{24}$

20. 30 days ÷ 5 = 6 days

6 days × 3 = 18 days

Lesson Test 5

1.
$$
\begin{array}{r}
{}^{6}\!\not{7}.{}^{1}25 \\
-\ 2.\ 8 \\
\hline
4.45
\end{array}
$$

2.
$$
\begin{array}{r}
{}^{4}\!\not{5}.{}^{1}07 \\
-\ 2.\ 11 \\
\hline
2.96
\end{array}
$$

3.
$$
\begin{array}{r}
{}^{5}\!\not{6}.{}^{1}\!\not{2}{}^{1}49 \\
-\ 5.\ 3\ 71 \\
\hline
0.\ 8\ 78
\end{array}
$$

4.
$$
\begin{array}{r}
{}^{1} \\
8.2 \\
+\ .9 \\
\hline
9.1
\end{array}
$$

5.
$$
\begin{array}{r}
6.14 \\
+\ .395 \\
\hline
6.535
\end{array}
$$

6.
$$
\begin{array}{r}
{}^{1} \\
14.9 \\
+\ .124 \\
\hline
15.024
\end{array}
$$

7. $\dfrac{2}{5} + \dfrac{1}{8} = \dfrac{16}{40} + \dfrac{5}{40} = \dfrac{21}{40}$

8. $\dfrac{3}{7} + \dfrac{2}{9} = \dfrac{27}{63} + \dfrac{14}{63} = \dfrac{41}{63}$

9. $\dfrac{2}{3} + \dfrac{1}{6} = \dfrac{12}{18} + \dfrac{3}{18} = \dfrac{15}{18} = \dfrac{5}{6}$

10. 29,000.1

11. 580.034

12. $\dfrac{3}{4} - \dfrac{1}{5} = \dfrac{15}{20} - \dfrac{4}{20} = \dfrac{11}{20}$

13. $\dfrac{7}{8} - \dfrac{2}{3} = \dfrac{21}{24} - \dfrac{16}{24} = \dfrac{5}{24}$

14. $\dfrac{1}{4} - \dfrac{1}{6} = \dfrac{6}{24} - \dfrac{4}{24} = \dfrac{2}{24} = \dfrac{1}{12}$

15. $20.00 - $15.24 = $4.76

16. $7.35 + $2.35 + $1.17 + $1.00 = $11.87

17. $35.00 - $11.87 = $23.13

18. 3.2 hr + 2.3 hr + 1.05 hr = 6.55 hr

19. 100 cents $\div$ 10 = 10 cents

10 cents $\times$ 2 = 20 cents

20. $\dfrac{3}{4} - \dfrac{1}{10} = \dfrac{30}{40} - \dfrac{4}{40} = \dfrac{26}{40} = \dfrac{13}{20}$

18. 1,000 g

19. 6.1 min $-$ 5.7 min = 0.4 min

20. $3 \times 10^3 + 4 \times 10^2 + 1 \times \dfrac{1}{10^1} + 2 \times \dfrac{1}{10^2}$

Lesson Test 6

1.

1 kilogram(kg)	1 hectogram(hg)
1,000 grams(g)	100 grams(g)
1 dekagram(dag)	1 gram(g)
10 grams(g)	1 gram(g)

2.

1 kiloliter(kl)	1 hectoliter(hl)
1,000 liters(L)	100 liters(L)
1 dekaliter(dal)	1 liter(L)
10 liters(L)	1 liter(L)

3.

1 kilometer(km)	1 hectometer(h
1,000 meters(m)	100 meters(m
1 dekameter(dam)	1 meter(m)
10 meters(m)	1 meter(m)

4. c

5. a

6. b

7. $\begin{array}{r} 1.2 \\ -\ .4 \\ \hline .8 \end{array}$

8. $\begin{array}{r} {}^4\!5\!.^1\!0 \\ -\ 2.\,8 \\ \hline 2.\,2 \end{array}$

9. $\begin{array}{r} 25.32 \\ +\ 1.06 \\ \hline 26.38 \end{array}$

10. $\begin{array}{r} 7.^3\!4^1\!2 \\ -2.\,3\ 9 \\ \hline 5.\,0\ 3 \end{array}$

11. $\dfrac{1}{5} + \dfrac{1}{9} = \dfrac{9}{45} + \dfrac{5}{45} = \dfrac{14}{45}$

12. $\dfrac{3}{8} - \dfrac{1}{4} = \dfrac{12}{32} - \dfrac{8}{32} = \dfrac{4}{32} = \dfrac{1}{8}$

13. $\dfrac{7}{9} - \dfrac{2}{7} = \dfrac{49}{63} - \dfrac{18}{63} = \dfrac{31}{63}$

14. $5\dfrac{2}{3} = \dfrac{15}{3} + \dfrac{2}{3} = \dfrac{17}{3}$

15. $2\dfrac{6}{7} = \dfrac{14}{7} + \dfrac{6}{7} = \dfrac{20}{7}$

16. $1\dfrac{5}{8} = \dfrac{8}{8} + \dfrac{5}{8} = \dfrac{13}{8}$

17. kilometers

Lesson Test 7

1.

1 gram(g)	1 decigram(dg)
1 gram(g)	$\dfrac{1}{10}$ gram(g)
1 centigram(cg)	1 milligram(mg)
$\dfrac{1}{100}$ gram(g)	$\dfrac{1}{1,000}$ gram(g)

2.

1 liter(L)	1 deciliter(dl)
1 liter(L)	$\dfrac{1}{10}$ liter(L)
1 centiliter(cl)	1 milliliter(ml)
$\dfrac{1}{100}$ liter(L)	$\dfrac{1}{1,000}$ liter(L)

3.

1 meter(m)	1 decimeter(dm)
1 meter(m)	$\dfrac{1}{10}$ meter(m)
1 centimeter(cm)	1 millimeter(mm)
$\dfrac{1}{100}$ meter(m)	$\dfrac{1}{1,000}$ meter(m)

4. kilo

5. hecto

6. deka

7. 40 $\div$ 2 = 20

20 $\times$ 1 = 20

8. 64 $\div$ 8 = 8

8 $\times$ 3 = 24

9. 14 $\div$ 7 = 2

2 $\times$ 2 = 4

10. $\dfrac{17}{8} = \dfrac{16}{8} + \dfrac{1}{8} = 2\dfrac{1}{8}$

11. $\dfrac{35}{4} = \dfrac{32}{4} + \dfrac{3}{4} = 8\dfrac{3}{4}$

12. $\dfrac{29}{3} = \dfrac{27}{3} + \dfrac{2}{3} = 9\dfrac{2}{3}$

13. Pippin

14. 5 quarts

15. kilograms

16. kiloliter

17. half a mile

18. two pounds

19. 1.2 in + 2.14 in + 1.75 in = 5.09 in

20. 50.9 in $-$ 5.09 in = 45.81 in

Lesson Test 8

1. 45,000 liters
2. 6,000 mg
3. 1,300 mm
4. b
5. a
6. c
7. e
8. f
9. d
10. c
11. a
12. b
13. $6\dfrac{3}{4}+3\dfrac{1}{5}=6\dfrac{15}{20}+3\dfrac{4}{20}=9\dfrac{19}{20}$
14. $11\dfrac{1}{9}+4\dfrac{2}{7}=11\dfrac{7}{63}+4\dfrac{18}{63}=15\dfrac{25}{63}$
15. $21\dfrac{2}{5}+7\dfrac{5}{6}=21\dfrac{12}{30}+7\dfrac{25}{30}=$
    $28\dfrac{37}{30}=29\dfrac{7}{30}$
16. They are the same.
17. 1 kilogram
18. 5 grams
19. 500 centigrams
20. $4\dfrac{1}{2}+4\dfrac{1}{2}=8\dfrac{2}{2}=9$ pies

Unit Test I

1. $1\times1\times1\times1\times1\times1=1$
2. $10\times10\times10=1,000$
3. $9\times9=81$
4. $3\times10+5\times1+2\times\dfrac{1}{10}+4\times\dfrac{1}{100}$
5. $9\times10^{4}+1\times\dfrac{1}{10^{1}}+6\times\dfrac{1}{10^{2}}$
6. 4,168.321
7. $\overset{5}{6}.\overset{1}{3}9$
   $-\ 3.54$
   $\overline{2.85}$
8. $\overset{4}{5}.\overset{9}{0}\overset{1}{0}$
   $-\ 2.15$
   $\overline{2.85}$

9. $1.\overset{7}{8}\overset{9}{0}\overset{1}{1}$
   $-\ .999$
   $\overline{.802}$
10. 8.3
    $+\ .4$
    $\overline{8.7}$
11. 1.23
    $+\ .147$
    $\overline{1.377}$
12. 91.2
    $+\ .608$
    $\overline{91.808}$
13. $96\div8=12$
    $12\times7=84$
14. $81\div9=9$
    $9\times5=45$
15. $\dfrac{6}{11}=\dfrac{12}{22}=\dfrac{18}{33}=\dfrac{24}{44}$
16. $\dfrac{9}{10}=\dfrac{18}{20}=\dfrac{27}{30}=\dfrac{36}{40}$
17. $\dfrac{1}{6}-\dfrac{1}{7}=\dfrac{7}{42}-\dfrac{6}{42}=\dfrac{1}{42}$
18. $\dfrac{3}{4}-\dfrac{2}{5}=\dfrac{15}{20}-\dfrac{8}{20}=\dfrac{7}{20}$
19. $\dfrac{7}{9}-\dfrac{4}{9}=\dfrac{3}{9}=\dfrac{1}{3}$
20. $1\dfrac{2}{3}+4\dfrac{1}{6}=1\dfrac{12}{18}+4\dfrac{3}{18}=5\dfrac{15}{18}=\boxed{5\dfrac{5}{6}}$
21. $10\dfrac{3}{9}+7\dfrac{1}{4}=$
    $10\dfrac{12}{36}+7\dfrac{9}{36}=17\dfrac{21}{36}=\boxed{17\dfrac{7}{12}}$
22. $34\dfrac{5}{6}+6\dfrac{1}{8}=$
    $34\dfrac{40}{48}+6\dfrac{6}{48}=40\dfrac{46}{48}=\boxed{40\dfrac{23}{24}}$
23. 1,200 cm
24. 900,000 cg
25. 220 ml
26. gram
27. kilogram
28. meter
29. 52.5 mi $+15.06$ mi $=67.56$ mi
30. $\$50.00-\$16.95=\$33.05$

Lesson Test 9

1.
$$\begin{array}{r} 1.3 \\ \times\ .2 \\ \hline .26 \end{array}$$

2.
$$\begin{array}{r} 2.3 \\ \times 1.2 \\ \hline 4\ 6 \\ 2\ 3 \\ \hline 2.7\ 6 \end{array}$$

3.
$$\begin{array}{r} .7 \\ \times\ .1 \\ \hline .07 \end{array}$$

4.
$$\begin{array}{r} 1.1 \\ \times\ .6 \\ \hline .66 \end{array}$$

5.
$$\begin{array}{r} 1.4 \\ \times 1.2 \\ \hline 2\ 8 \\ 1\ 4 \\ \hline 1.6\ 8 \end{array}$$

6.
$$\begin{array}{r} 2.0 \\ \times\ .4 \\ \hline .80 \end{array}$$

7. 250 hg

8. 800 cm

9. 310 ml

10. liter

11. kilometer

12. kilogram

13. $12\frac{1}{4} - 3\frac{1}{5} = 12\frac{5}{20} - 3\frac{4}{20} = 9\frac{1}{20}$

14. $9 - 2\frac{7}{8} = 8\frac{8}{8} - 2\frac{7}{8} = 6\frac{1}{8}$

15. $21\frac{1}{3} - 6\frac{3}{4} = 21\frac{4}{12} - 6\frac{9}{12} =$
$20\frac{16}{12} - 6\frac{9}{12} = 14\frac{7}{12}$

16. 10 grams

17. 4.2 mi $\times$ 0.2 = ~~0.084 mi~~ .84

18. 3.3 lb + 2.75 lb + 4.09 lb = 10.14 lb

19. $4 - 1\frac{5}{6} = 3\frac{6}{6} - 1\frac{5}{6} = 2\frac{1}{6}$ pies

20. 25.6 mi − 19.8 mi = 5.8 mi

Lesson Test 10

1.
$$\begin{array}{r} 6.24 \\ \times\ .4 \\ \hline 2.496 \end{array}$$

2.
$$\begin{array}{r} 4.4 \\ \times 1.7 \\ \hline 3.08 \\ 4.4 \\ \hline 7.48 \end{array}$$

3.
$$\begin{array}{r} .67 \\ \times .12 \\ \hline .0134 \\ .067 \\ \hline .0804 \end{array}$$

4.
$$\begin{array}{r} 7.18 \\ \times\ .7 \\ \hline 5.026 \end{array}$$

5.
$$\begin{array}{r} 3.3 \\ \times 1.5 \\ \hline 1.65 \\ 3.3 \\ \hline 4.95 \end{array}$$

6.
$$\begin{array}{r} .85 \\ \times .01 \\ \hline .0085 \end{array}$$

7. 900,000 cg

8. 48,000 mm

9. 12.17 + 147.09 = 159.26

10. 5.13 + 0.26 = 5.39

11. 18.7 − 4.8 = 13.9

12. $4\frac{3}{4} + 1\frac{1}{10} = 4\frac{30}{40} + 1\frac{4}{40} =$
$5\frac{34}{40} = 5\frac{17}{20}$

13. $13\frac{1}{2} - 4\frac{1}{8} = 13\frac{8}{16} - 4\frac{2}{16} =$
$9\frac{6}{16} = 9\frac{3}{8}$

14. $20 - 10\frac{1}{5} = 19\frac{5}{5} - 10\frac{1}{5} = 9\frac{4}{5}$

15. $\frac{5}{6} \times \frac{1}{4} = \frac{5}{24}$

16. $\frac{\cancel{3}}{\cancel{8}_4} \times \frac{\cancel{2}}{\cancel{3}} = \frac{1}{4}$

17. $\frac{5}{\cancel{6}_3} \times \frac{\cancel{2}}{3} = \frac{5}{9}$

18. $4 \times \$9.24 = \$36.96$
19. $642 \text{ km} \times 0.6 \text{ mi/km} = 385.2 \text{ mi}$
20. $\$59.50 \times 1.75 \approx \$104.13$

Lesson Test 11

1. 0.35
2. 0.06
3. 0.19
4. 0.15
5. 0.58
6. 0.07
7. $\dfrac{2}{100} = \dfrac{1}{50}$
8. $\dfrac{23}{100}$
9. $\dfrac{44}{100} = \dfrac{11}{25}$
10. $1 \div 5 = 0.20 = 20\%$
11. $1 \div 2 = 0.50 = 50\%$
12. $1 \div 4 = 0.25 = 25\%$
13. $3 \div 4 = 0.75 = 75\%$
14.
$$\begin{array}{r} 9.30 \\ \times\ .01 \\ \hline 930 \\ 000 \\ \hline .0930 \end{array}$$
15.
$$\begin{array}{r} 3.5 \\ \times 4.5 \\ \hline 2 \\ 155 \\ 2 \\ 120 \\ \hline 15.75 \end{array}$$
16.
$$\begin{array}{r} 2.56 \\ \times\ .81 \\ \hline 256 \\ 608 \\ \hline 2.0736 \end{array}$$
17. $\dfrac{\overset{15}{\cancel{30}}}{7} \times \dfrac{3}{\cancel{2}} = \dfrac{45}{7} = 6\dfrac{3}{7}$
18. $\dfrac{\cancel{5}}{3} \times \dfrac{34}{\cancel{5}} = \dfrac{34}{3} = 11\dfrac{1}{3}$
19. $\$15.96 \times 0.15 = \$2.394$
    $\$15.96 + 2.39 = \$18.35$
20. $20 \text{ km} \times 0.60 \text{ mi/km} = 12 \text{ mi}$

Lesson Test 12

1. $\dfrac{300}{100} + \dfrac{75}{100} = \dfrac{375}{100} = 375\% = 3.75$
2. $\dfrac{100}{100} + \dfrac{40}{100} = \dfrac{140}{100} = 140\% = 1.40$
3. $\dfrac{600}{100} + \dfrac{90}{100} = \dfrac{690}{100} = 690\% = 6.90$
4. $\dfrac{200}{100} + \dfrac{50}{100} = \dfrac{250}{100} = 250\% = 2.50$
5. $0.35 = \dfrac{35}{100} = \dfrac{7}{20}$
6. $0.11 = \dfrac{11}{100}$
7. yard
8. ounce
9. inch
10. inch
11. pounds
12. mile
13. quarts
14. $\dfrac{3}{5} \div \dfrac{1}{3} = \dfrac{9}{15} \div \dfrac{5}{15} = \dfrac{9 \div 5}{1} = \dfrac{9}{5} = 1\dfrac{4}{5}$
15. $\dfrac{3}{4} \div \dfrac{7}{8} = \dfrac{24}{32} \div \dfrac{28}{32} = \dfrac{24 \div 28}{1} = \dfrac{24}{28} = \dfrac{6}{7}$
16. $\dfrac{2}{5} \div \dfrac{3}{8} = \dfrac{16}{40} \div \dfrac{15}{40} = \dfrac{16 \div 15}{1} = \dfrac{16}{15} = 1\dfrac{1}{15}$
17. $\$11.25 \times 1.10 = \$12.38 \text{ (rounded)}$
18. $25 \times 4.00 = 100 \text{ cards}$
19. $\dfrac{1}{2} \div \dfrac{1}{10} = \dfrac{10}{20} \div \dfrac{2}{20} = \dfrac{10 \div 2}{1} = 5 \text{ guests}$
20. $\$75.00 \times 0.25 = \$18.75$

Lesson Test 13

1.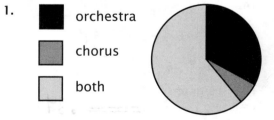
   - orchestra
   - chorus
   - both
2. $2 \text{ hr} \times 60 \text{ min/hr} = 120 \text{ min}$
   $120 \text{ min} \times 0.05 = 6 \text{ minutes}$
3. cloudy
4. $30 \text{ days} \times 0.30 = 9 \text{ days}$
5. $30 \text{ days} \times 0.40 = 12 \text{ days}$
6. $\dfrac{400}{100} + \dfrac{75}{100} = \dfrac{475}{100} = 475\% = 4.75$

7. 0.25

8. 0.16

9. 0.06

10. 1.25

11. $3.61 \times 0.07 = 0.2527$

12. $0.9 - 0.14 = 0.76$

13. $2.69 + 8.7 = 11.39$

14. $\dfrac{44}{5} \div \dfrac{14}{3} = \dfrac{132}{15} \div \dfrac{70}{15} = \dfrac{132 \div 70}{1}$

$= \dfrac{132}{70} = \dfrac{66}{35} = 1\dfrac{31}{35}$

15. $\dfrac{17}{5} \div \dfrac{47}{7} = \dfrac{119}{35} \div \dfrac{235}{35} = \dfrac{119}{235}$

16. $\dfrac{29}{2} \div \dfrac{17}{3} = \dfrac{87}{6} \div \dfrac{34}{6} = \dfrac{87}{34} = 2\dfrac{19}{34}$

17. $\$1.45 \times 1.15 = \$1.67$ (rounded)

18. $10 \text{ kg} \times 1{,}000 \text{ g/kg} = 10{,}000$ grams

19. $1.5 \text{ books/week} \times 6 \text{ weeks} = 9$ books

20. A liter is close to 1 quart,
so approximately 4 liters.

Lesson Test 14

1. $(0.1) \times (0.5) = (0.05)$

$$
\begin{array}{r}
.123 \\
\times \quad .456 \\
\hline
11 \\
628 \\
{}^1 15 \\
50 \\
12 \\
{}^1 48 \\
\hline
.056088
\end{array}
$$

2. $(3) \times (6) = (18)$

$$
\begin{array}{r}
3.052 \\
\times \quad 6.193 \\
\hline
{}^1 9\ {}^1 156 \\
{}^1 2\ {}^2 7\ 4\ 1 \\
58 \\
3\ 0\ 5\ 2 \\
18\ 3\ 12 \\
\hline
18.9\ 0\ 1\ 036
\end{array}
$$

3. $(0.3) \times (0.3) = (0.09)$

$$
\begin{array}{r}
.281 \\
\times \quad .269 \\
\hline
729 \\
{}^2 18 \\
{}^1 486 \\
2 \\
16\ 2 \\
4 \\
\hline
.075589
\end{array}
$$

4. $(2) \times (5) = (10)$

$$
\begin{array}{r}
1.504 \\
\times \quad 5.167 \\
\hline
{}^1 3\ 528 \\
7 \\
{}^1 3\ 0\ 24 \\
6 \\
{}^1 5\ 0\ 4 \\
2\ 5\ 20 \\
5 \\
\hline
7.77\ 1\ 168
\end{array}
$$

5. 5

6. $\dfrac{1}{2}$

7. $\dfrac{38}{100} = \dfrac{19}{50}$

8. $\dfrac{2}{100} = \dfrac{1}{50}$

9. $\dfrac{900}{100} + \dfrac{40}{100} = \dfrac{940}{100} = 940\% = 9.40$

10. 16,000 mm

11. 350 hl

12. 10 dg

13. $\dfrac{5}{8} \div \dfrac{2}{3} = \dfrac{5}{8} \times \dfrac{3}{2} = \dfrac{15}{16}$

14. $\dfrac{1}{5} \div \dfrac{3}{8} = \dfrac{1}{5} \times \dfrac{8}{3} = \dfrac{8}{15}$

15. $\dfrac{1}{2} \div \dfrac{3}{5} = \dfrac{1}{2} \times \dfrac{5}{3} = \dfrac{5}{6}$

16. $10.4 \text{ gal} \times \$1.639\text{/gal} = \$17.05$

17. $3.2 \text{ m} \times 39.37 \text{ in/m} = 125.984$ in

18. $\$6.00 \times 1.20 = \$7.20$

19. $\$25.60 + \$11.19 + \$45.21 = \$82.00$

20. $\$82.00 \times 1.11 = \$91.02$

Lesson Test 15

1. 0.07 m
2. 0.1 kl
3. 0.5 g
4. 0.25 hm
5. 80 ml
6. 9,500 cg
7. $(1) \times (0.007) = (0.007)$

$$
\begin{array}{r}
1.465 \\
\times \quad .007 \\
\hline
{}^1 2\,4\,3 \\
7\,8\,2\,5 \\
\hline
0.010\,2\,5\,5
\end{array}
$$

8. $(200) \times (0.1) = (20)$

$$
\begin{array}{r}
239.016 \\
\times \quad .134 \\
\hline
{}^1 1\,3 \quad 2 \\
8\,2\,{}^1 6\,{}^1 0\,4\,4 \\
{}^3 6\,{}^2 9\,7\,0\,{}^1 3\,8 \\
{}^1 2\,3\,9\,0\,1\,6 \\
\hline
32.028\,1\,4\,4
\end{array}
$$

9. 0.06
10. 0.45
11. 1.30
12. 2.00
13. $34 \div 2 = 17$
14. $36 \div 9 = 4$
    $4 \times 2 = 8$
15. $100 \div 5 = 20$
    $20 \times 3 = 60$
16. $\dfrac{33}{8} \div \dfrac{6}{5} = \dfrac{\overset{11}{\cancel{33}}}{8} \times \dfrac{5}{\underset{2}{\cancel{6}}} = \dfrac{55}{16} = 3\dfrac{7}{16}$
17. $\dfrac{26}{3} \div \dfrac{1}{6} = \dfrac{26}{\cancel{3}} \times \dfrac{\overset{2}{\cancel{6}}}{1} = 52$
18. 1 kilogram
19. 500,000 mm
20. $35 \text{ tons} \times 1.08 = 37.8 \text{ tons}$

Lesson Test 16

1. $3.14(10 \text{ m})^2 = 314 \text{ m}^2$
2. $3.14(20 \text{ m}) = 62.8 \text{ m}$
3. $3.14(2.5 \text{ in})^2 = 19.625 \text{ in}^2$

4. $3.14(5 \text{ in}) = 15.7 \text{ in}$
5. 0.061 m
6. 0.004 kg
7. 130 ml
8. $0.07 \times 0.04 = 0.0028$
9. $72.3 \times 0.9 = 65.07$
10. $204 \times 0.11 = 22.44$
11. $\dfrac{17}{2} \div \dfrac{9}{4} = \dfrac{17}{\cancel{2}} \times \dfrac{\overset{2}{\cancel{4}}}{9} = \dfrac{34}{9} = 3\dfrac{7}{9}$
12. $\dfrac{19}{6} \div \dfrac{5}{3} = \dfrac{19}{\underset{2}{\cancel{6}}} \times \dfrac{\cancel{3}}{5} = \dfrac{19}{10} = 1\dfrac{9}{10}$
13. $\dfrac{9}{10} \div \dfrac{2}{5} = \dfrac{9}{\underset{2}{\cancel{10}}} \times \dfrac{\cancel{5}}{2} = \dfrac{9}{4} = 2\dfrac{1}{4}$
14. $9 \text{ m} + 4 \text{ m} + 9 \text{ m} + 4 \text{ m} = 26 \text{ m}$
15. $4.5 \text{ ft} + 4.5 \text{ ft} + 4.5 \text{ ft} + 4.5 \text{ ft} = 18 \text{ ft}$
16. $6.2 \text{ in} + 3 \text{ in} + 6.2 \text{ in} + 3 \text{ in} = 18.4 \text{ in}$
17. $3.14(6 \text{ ft})^2 = 113.04 \text{ ft}^2$
18. $3.14(5 \text{ ft}) = 15.7 \text{ ft}$
19. $8 \text{ yd} + 8 \text{ yd} + 8 \text{ yd} + 8 \text{ yd} = 32 \text{ yd}$
20. $2 \text{ hm} \times 100 \text{ m/hm} = 200 \text{ m}$

Unit Test II

1.
$$
\begin{array}{r}
8.6\,1 \\
\times \quad .7 \\
\hline
1\,4 \\
5\,6\,2\,7 \\
\hline
6.0\,2\,7
\end{array}
$$

2.
$$
\begin{array}{r}
1.4 \\
\times 2.6 \\
\hline
2 \\
1\,6\,4 \\
2\,8 \\
\hline
3.6\,4
\end{array}
$$

3.
$$
\begin{array}{r}
.18 \\
\times .95 \\
\hline
1\,4 \\
7\,5\,0 \\
9\,2 \\
\hline
.17\,1\,0
\end{array}
$$

4.
$$
\begin{array}{r}
.106 \\
\times\ .352 \\
\hline
2\,1\,2 \\
{}^{1}5\,3\,0 \\
3\,1\,8 \\
\hline
.0\,3\,7\,3\,1\,2
\end{array}
$$

5.
$$
\begin{array}{r}
2.605 \\
\times\ 6.279 \\
\hline
5 \\
{}^{1}1\,8\,4\,4\,5 \\
4 \\
{}^{1}1\,4\,2\,3\,5 \\
1 \\
{}^{1}3\,4\,2\,1\,0 \\
1\,2\,6\,3\,0 \\
\hline
1\,6.3\,5\,6\,7\,9\,5
\end{array}
$$

6.
$$
\begin{array}{r}
3.191 \\
\times\ 4.260 \\
\hline
5\quad 0 \\
{}^{1}{}^{2}8\,6\,4\,6 \\
1 \\
6\,2\,8\,2 \\
3 \\
1\,{}^{1}2\,4\,6\,4 \\
\hline
1\,3.5\,9\,3\,6\,6\,0
\end{array}
$$

7. 0.01

8. 0.10

9. 1.00

10. $\dfrac{5}{100} = \dfrac{1}{20}$

11. $\dfrac{48}{100} = \dfrac{12}{25}$

12. $\dfrac{16}{100} = \dfrac{4}{25}$

13. 0.25 = 25%

14. 0.40 = 40%

15. 0.50 = 50%

16. 0.75 = 75%

17. $\dfrac{600}{100} + \dfrac{25}{100} = \dfrac{625}{100} = 625\% = 6.25$

18. 0.2 m

19. 8,100,000 cg

20.

■ electricity

▨ ingredients

▦ paper cups

21. $\$35 \times 0.80 = \$28$

22. $3.14(3\ \text{m})^2 = 28.26\ \text{m}^2$

23. $3.14(6\ \text{m}) = 18.84\ \text{m}$

24. $\dfrac{27}{5} \times \dfrac{21}{5} = \dfrac{567}{25} = 22\dfrac{17}{25}$

25. $\dfrac{7}{3} \times \dfrac{5}{9} = \dfrac{35}{27} = 1\dfrac{8}{27}$

26. $\dfrac{15}{2} \div \dfrac{5}{4} = \dfrac{{}^{3}15}{2} \times \dfrac{4^{2}}{5} = 6$

27. $\dfrac{20}{3} \div \dfrac{2}{5} = \dfrac{{}^{10}20}{3} \times \dfrac{5}{2} = \dfrac{50}{3} = 16\dfrac{2}{3}$

28. $0.88 \times 50 = 44$ questions

29. $\$135.00 \times 1.16 = \$156.60$

30. $10.16 + 9.25 + 10.16 + 9.25 = 38.82$ in

Lesson Test 17

1.
$$
\begin{array}{r}
.002 \\
2\overline{)\,.004} \\
\underline{4} \\
0
\end{array}
$$

2.
$$
\begin{array}{r}
.002 \\
\times\quad 2 \\
\hline
.004
\end{array}
$$

3.
$$
\begin{array}{r}
.22 \\
8\overline{)\,1.76} \\
\underline{160} \\
16 \\
\underline{16} \\
0
\end{array}
$$

4.
$$
\begin{array}{r}
.22 \\
\times\quad 8 \\
\hline
1 \\
1\ 6\ 6 \\
\hline
1.7\ 6
\end{array}
$$

Lesson Test 18

5.
```
    .07
  4⌐.28
    28
     0
```

6.
```
   .07
 ×   4
   .28
```

7.
```
  0.2
5⌐1.0
   10
    0
```

8.
```
   .2
 × 5
  1.0
```

9.
```
  0.5
7⌐3.5
   35
    0
```

10.
```
   .5
 × 7
  3.5
```

11.
```
   .003
 3⌐.009
     9
     0
```

12.
```
   .003
 ×    3
   .009
```

13. $3.14(1.2 \text{ mi})^2 = 4.5216 \text{ mi}^2$

14. $3.14(2.4 \text{ mi}) = 7.536 \text{ mi}$

15. $40 \text{ in} + 30 \text{ in} + 40 \text{ in} + 30 \text{ in} = 140 \text{ in}$

16. $7.9 \text{ m} + 3.2 \text{ m} + 7.9 \text{ m} + 3.2 \text{ m} = 22.2 \text{ m}$

17. $6\frac{1}{4} + 4\frac{3}{4} + 6\frac{1}{4} + 4\frac{3}{4} = 20\frac{8}{4} = 22 \text{ in}$

18. $\$40.95 \div 7 = \$5.85$

19. $71.6 \text{ mi} \div 4 = 17.9 \text{ mi}$

20. $3.5 \text{ hr} + 6.9 \text{ hr} + 4.3 \text{ hr} = 14.7 \text{ hr}$
$14.7 \text{ hr} \times \$7.10/\text{hr} = \$104.37$

1.
```
          121,000
  .004⌐484.000
        400,000
         84,000
         80,000
          4,000
          4,000
              0
```

2.
```
   121,000
 ×    .004
   484.000
```

3.
```
       201
 .18⌐36.18
      3600
        18
        18
         0
```

4.
```
     201
 ×   .18
    1608
     201
   36.18
```

5.
```
    3.7
 6⌐22.2
   18.0
     42
     42
      0
```

6.
```
   3.7
 ×   6
    14
   182
  22.2
```

7.
```
    .0006
 9⌐.0054
      54
       0
```

8.
```
   .0006
 ×     9
   .0054
```

9. $3.14(3.5 \text{ in})^2 = 38.465 \text{ in}^2$

10. $3.14(7 \text{ in}) = 21.98 \text{ in}$

11. $13 \times 13 = 169$
12. $6 \times 6 = 36$
13. $10 \times 10 = 100$
14. $11 \text{ ft} + 11 \text{ ft} + 12 \text{ ft} = 34 \text{ ft}$
15. $9 \text{ in} + 12 \text{ in} + 15 \text{ in} = 36 \text{ in}$
16. $6.9 \text{ m} + 5.4 \text{ m} + 4.1 \text{ m} = 16.4 \text{ m}$
17. $\$19 \div 0.25 = 76 \text{ quarters}$
18. $5.4 \text{ lb} \div 0.45 \text{ lb/batch} = 12 \text{ batches}$
19. $13.9 + 13.9 + 13.9 + 13.9 = 55.6 \text{ m}$
20. $3{,}160 \text{ g} \times \dfrac{1 \text{ kg}}{1{,}000 \text{ g}} = 3.16 \text{ kg}$

Lesson Test 19

1. $0.19G = 38$
   $G = 38 \div 0.19$
   $G = 200$

2. $0.19(200) = 38$
   $38 = 38$

3. $0.008Y = 2$
   $Y = 2 \div 0.008$
   $Y = 250$

4. $0.008(250) = 2$
   $2 = 2$

5. $0.6G = 24$
   $G = 24 \div 0.6$
   $G = 40$

6. $0.6(40) = 24$
   $24 = 24$

7. $0.11Y = 55$
   $Y = 55 \div 0.11$
   $Y = 500$

8. $0.11(500) = 55$
   $55 = 55$

9.
```
        1,003
006. 6 018.
      6,000
         18
         18
          0
```

10.
```
    1,003
×   .006
    6.018
```

11.
```
       0.2
15 3.0
      30
       0
```

12.
```
     15
×    .2
    3.0
```

13. $8 \text{ m} \times 4 \text{ m} = 32 \text{ m}^2$
14. $3.9 \text{ ft} \times 3.9 \text{ ft} = 15.21 \text{ ft}^2$
15. $5.1 \text{ in} \times 2.5 \text{ in} = 12.75 \text{ in}^2$
16. $0.45J = 90 \text{ hr}$
    $J = 90 \text{ hr} \div 0.45$
    $J = 200 \text{ hours}$
17. $200 \text{ hr} - 90 \text{ hr} = 110 \text{ hr}$
18. $0.75G = 12$
    $G = 12 \div 0.75$
    $G = 16 \text{ gifts}$
19. $\$308.00 \div 8 \text{ people} = \$38.50/\text{person}$
20. $5{,}000 \text{ g} = 5 \text{ kg}$
    $50 \text{ kg} > 5 \text{ kg}$

Lesson Test 20

1.
```
          55
5. 275.
       250
        25
        25
         0
```

2.
```
     5 5
×    .5
      2
   25 5
   27.5
```

3.
```
         21.25
32. 680.00
      640.00
       40.00
       32.00
        8.00
        6.40
        1.60
        1.60
           0
```

4.
$$
\begin{array}{r}
21.25 \\
\times \quad .32 \\
\hline
1\;1\;1 \\
4\;2\;4\;0 \\
1 \\
6\;3\;6\;5 \\
\hline
6.8\;0\;0\;0
\end{array}
$$

5.
$$
\begin{array}{r}
680 \\
04.\overline{)2720.} \\
\underline{2400} \\
320 \\
\underline{320} \\
0
\end{array}
$$

6.
$$
\begin{array}{r}
6\;8\;0 \\
\times \quad .04 \\
\hline
3 \\
2\;4\;2\;0 \\
\hline
27.2\;0
\end{array}
$$

7.
$$
\begin{array}{r}
15 \\
28.\overline{)420.} \\
\underline{280} \\
140 \\
\underline{140} \\
0
\end{array}
$$

8.
$$
\begin{array}{r}
.28 \\
\times 15 \\
\hline
1 \\
140 \\
28 \\
\hline
4.20
\end{array}
$$

9.
$$
\begin{array}{r}
.056 \\
35\overline{)1.960} \\
\underline{1750} \\
210 \\
\underline{210} \\
0
\end{array}
$$

10.
$$
\begin{array}{r}
.056 \\
\times \quad 35 \\
\hline
1\;3 \\
2\;5\;0 \\
1 \\
1\;5\;8 \\
\hline
1.9\;6\;0
\end{array}
$$

11.
$$
\begin{array}{r}
4,270 \\
002.\overline{)8,540.} \\
\underline{8,000} \\
540 \\
\underline{400} \\
140 \\
\underline{140} \\
0
\end{array}
$$

12.
$$
\begin{array}{r}
4\;2\;7\;0 \\
\times \quad .002 \\
\hline
1 \\
8\;4\;4\;0 \\
\hline
8.5\;4\;0
\end{array}
$$

13. $0.12Q = 0.36$
$Q = 0.36 \div 0.12$
$Q = 3$
$0.12(3) = 0.36;\ 0.36 = 0.36$

14. $2.5W = 0.75$
$W = 0.75 \div 2.5$
$W = 0.3$
$2.5(0.3) = 0.75;\ 0.75 = 0.75$

15. $\dfrac{17}{3} \div \dfrac{6}{7} = \dfrac{17}{3} \times \dfrac{7}{6} = \dfrac{119}{18} = 6\dfrac{11}{18}$

16. $23\dfrac{10}{10} - 12\dfrac{1}{10} = 11\dfrac{9}{10}$

17. $\dfrac{4}{\cancel{5}} \times \dfrac{\cancel{10}^{2}}{11} = \dfrac{8}{11}$

18. $10\ \text{in} \times 6\ \text{in} = 60\ \text{in}^2$

19. $27.3\ \text{mi} \div 9.1\ \text{mi/part} = 3\ \text{parts}$

20. $\$66.36 \div \$3.16/\text{item} = 21\ \text{items}$

Lesson Test 21

1.
$$
\begin{array}{r}
21.325 = 21.33 \\
4.\overline{)8,5.300} \\
\underline{8,0000} \\
5300 \\
\underline{4000} \\
1300 \\
\underline{1200} \\
100 \\
\underline{80} \\
20
\end{array}
$$

2.
$$
\begin{array}{r}
.245 = 0.25 \\
16\overline{)3.930} \\
\underline{3.200} \\
730 \\
\underline{640} \\
90 \\
\underline{80} \\
10
\end{array}
$$

3.
$$
\begin{array}{r}
2.066 = 2.07 \\
09.\overline{)18.600} \\
\underline{18000} \\
600 \\
\underline{540} \\
60 \\
\underline{54} \\
6
\end{array}
$$

4.
$$
\begin{array}{r}
7.4545 = 7.\overline{45} \\
11.\overline{)82.0000} \\
\underline{770000} \\
50000 \\
\underline{44000} \\
6000 \\
\underline{5500} \\
500 \\
\underline{440} \\
60 \\
\underline{55} \\
5
\end{array}
$$

5.
$$
\begin{array}{r}
44.0909 = 44.\overline{09} \\
22.\overline{)970.0000} \\
\underline{8800000} \\
900000 \\
\underline{880000} \\
20000 \\
\underline{19800} \\
200 \\
\underline{198} \\
2
\end{array}
$$

6.
$$
\begin{array}{r}
1.166 = 1.1\overline{6} \\
6\overline{)7.000} \\
\underline{6000} \\
1000 \\
\underline{600} \\
400 \\
\underline{360} \\
40 \\
\underline{36} \\
4
\end{array}
$$

7.
$$
30X = 0.8 \\
X = 0.8 \div 30 \\
X = 0.02\tfrac{2}{3}
$$

8.
$$
0.5Y = 0.123 \\
Y = 0.123 \div 0.5 \\
Y = 0.24\tfrac{3}{5}
$$

9.
$$
2.3F = 6.41 \\
F = 6.41 \div 2.3 \\
F = 2.78\tfrac{16}{23}
$$

10. $65 - 0.45 = 64.55$
11. $7.03 + 0.2 = 7.23$
12. $200 - 0.02 = 199.98$
13. $1/2(12 \text{ ft} \times 18 \text{ ft}) = 108 \text{ ft}^2$
14. $1/2(5 \text{ in} \times 12 \text{ in}) = 30 \text{ in}^2$
15. $1/2(10.2 \text{ m} \times 4.1 \text{ m}) = 20.91 \text{ m}^2$
16. $51.84 \text{ m}^2 \div 5.4 \text{ m} = 9.6 \text{ m}$
17. $\$25.45 \times 1.20 = \$30.54$
18. $3.14(8 \text{ in})^2 = 200.96 \text{ in}^2$
19. grams
20. $31 \text{ kg} \times 1,000 \text{ g/kg} = 31,000 \text{ g}$
    $31,000 \text{ g} - 310 \text{ g} = 30,690 \text{ g}$

Lesson Test 22

1. $0.09G + 0.38 = 0.425$
   $0.09G = 0.425 - 0.38$
   $0.09G = 0.045$
   $G = 0.045 \div 0.09 = 0.5$

2. $0.09(0.5) + 0.38 = 0.425$
   $0.045 + 0.38 = 0.425$
   $0.425 = 0.425$

3. $0.4X + 3.6 = 6.4$
$0.4X = 6.4 - 3.6$
$0.4X = 2.8$
$X = 2.8 \div 0.4 = 7$

4. $0.4(7) + 3.6 = 6.4$
$2.8 + 3.6 = 6.4$
$6.4 = 6.4$

5. $1.5X + 4 = 4.165$
$1.5X = 4.165 - 4$
$1.5X = 0.165$
$X = 0.165 \div 1.5 = 0.11$

6. $1.5(.11) + 4 = 4.165$
$0.165 + 4 = 4.165$
$4.165 = 4.165$

7. $.944 = 0.94$

$$9. \overline{)8.500}$$
$$8100$$
$$400$$
$$360$$
$$40$$
$$36$$
$$4$$

8. $4.032 = 4.03$
$$4\overline{)16.130}$$
$$16000$$
$$130$$
$$120$$
$$10$$
$$8$$
$$2$$

9. $8.971 = 8.97$
$$07.\overline{)62.800}$$
$$56000$$
$$6800$$
$$6300$$
$$500$$
$$490$$
$$10$$
$$7$$
$$3$$

10. 1.8 m

11. 0.549 kl

12. 45 mg

13. $2.8 \text{ ft} \times 2.8 \text{ ft} \times 2.8 \text{ ft} = 21.952 \text{ ft}^3$

14. $\dfrac{5}{4} m \times \dfrac{5}{4} m \times \dfrac{5}{4} m = \dfrac{125}{64} \text{ m}^3 = 1\dfrac{61}{64} \text{ m}^3$

15. $3 \text{ in} \times 22 \text{ in} = 66 \text{ in}^2$

16. $22 \text{ in} + 4 \text{ in} + 22 \text{ in} + 4 \text{ in} = 52 \text{ in}$

17. $\dfrac{1}{2}(4 \text{ m} \times 6 \text{ m}) = 12 \text{ m}^2$

18. $5.4 \text{ m} + 4.2 \text{ m} + 6 \text{ m} = 15.6 \text{ m}$

19. $\$1.69 \times 6 = \$10.14$
$\$10.14 \times 1.03 = \$10.44$

20. $0.45D + 1.35 = 23.85$
$0.45D = 23.85 - 1.35$
$0.45D = 22.50$
$D = 22.50 \div 0.45 = \$50$

Lesson Test 23

1. $3 \div 10 = 0.30 = 30\%$

2. $1 \div 7 = 0.14\dfrac{2}{7} = 14\dfrac{2}{7}\%$

3. $2 \div 3 = 0.66\dfrac{2}{3} = 66\dfrac{2}{3}\%$

4. $5 \div 6 = 0.83\dfrac{1}{3} = 83\dfrac{1}{3}\%$

5. $1 \div 13 = 0.07\dfrac{9}{13} = 7\dfrac{9}{13}\%$

6. $8 \div 9 = 0.88\dfrac{8}{9} = 88\dfrac{8}{9}\%$

7. $0.9R + 0.5 = 0.77$
$0.9R = 0.77 - 0.5$
$0.9R = 0.27$
$R = 0.27 \div 0.9 = 0.3$

8. $0.9(0.3) + 0.5 = 0.77$
$0.27 + 0.5 = 0.77$
$0.77 = 0.77$

9. $5.2833 = 5.28\overline{3}$
$$6\overline{)31.7000}$$
$$300000$$
$$17000$$
$$12000$$
$$5000$$
$$4800$$
$$200$$
$$180$$
$$2$$

10. $303.33 = 303.\overline{3}$
3.$\overline{)910.00}$
90000
1000
900
100
90
10
9
1

11. $4.722 = 4.7\overline{2}$
09.$\overline{)42.500}$
36000
6500
6300
200
180
20
18
2

12. $8 \text{ in} \times 9 \text{ in} \times 5 \text{ in} = 360 \text{ in}^3$

13. $3.4\text{m} \times 2.1\text{m} \times 1\text{m} = 7.14 \text{ m}^3$

14. $\frac{1}{2} \text{ ft} \times \frac{1}{2} \text{ ft} \times \frac{3}{4} \text{ ft} = \frac{3}{16} \text{ ft}^3$

15. $\frac{35}{50} = \frac{70}{100} = 70\%$

16. $1 \div 8 = 0.12\frac{1}{2} = 12\frac{1}{2}\%$

17. $\$5.65/\text{lb} \times 0.60 \text{ lb} = \$3.39$

18. $45 \times .80 = 36$ done
$45 - 36 = 9$ toys left

19. $\$11.20 \times 1.20 = \$13.44$
$\$20.00 - \$13.44 = \$6.56$

20. $100 \text{ ft}^2 \div 10 \text{ ft} = 10 \text{ ft}$

Unit Test III

1. .002
4$\overline{).008}$
8
0

2. .002
× 4
.008

3. .05
6$\overline{).30}$
30
0

4. .05
× 6
.30

5. 430
2.$\overline{)860.}$
800
60
60
0

6. 430
× .2
86.0

7. .07
09.$\overline{)00.63}$
63
0

8. .07
× .09
.0063

9. 16
21.$\overline{)336.}$
210
126
126
0

10. 16
×2.1
1 16
2 2
33.6

11. 1,810
005.$\overline{)9,050.}$
5,000
4,050
4,000
50
50
0

12.
$$
\begin{array}{r}
1,8\,10 \\
\times \quad .005 \\
\hline
4 \\
5\,0\,5\,0 \\
\hline
9.0\,5\,0
\end{array}
$$

13. $6.3W = 11.34$
$\quad W = 11.34 \div 6.3$
$\quad W = 1.8$

14. $6.3(1.8) = 11.34$
$\quad 11.34 = 11.34$

15. $0.7X + 0.08 = 17.58$
$\quad 0.7X = 17.58 - 0.08$
$\quad 0.7X = 17.5$
$\quad X = 17.5 \div 0.7 = 25$

16. $0.7(25) + 0.08 = 17.58$
$\quad 17.5 + 0.08 = 17.58$
$\quad 17.58 = 17.58$

17. $2 \text{ in} \times 10 \text{ in} = 20 \text{ in}^2$

18. $10 \text{ in} + 5.5 \text{ in} + 10 \text{ in} + 5.5 \text{ in} = 31 \text{ in}$

19. $1.4 \text{ m} \times 0.7 \text{ m} = 0.98 \text{ m}^2$

20. $1.4 \text{ m} + 0.7 \text{ m} + 1.4 \text{ m} + 0.7 \text{ m} = 4.2 \text{ m}$

21. $\dfrac{1}{2}(3 \text{ ft} \times 6 \text{ ft}) = 9 \text{ ft}^2$

22. $2 \text{ ft} + 4 \text{ ft} + 6 \text{ ft} = 12 \text{ ft}$

23.
$$
\begin{array}{r}
12.233 \approx 12.23 \\
6\overline{)73.400} \\
\underline{60000} \\
13400 \\
\underline{12000} \\
1400 \\
\underline{1200} \\
200 \\
\underline{180} \\
20
\end{array}
$$

24.
$$
\begin{array}{r}
15.166 \approx 15.1\overline{6} \\
3.\overline{)45.500} \\
\underline{30000} \\
15500 \\
\underline{15000} \\
500 \\
\underline{300} \\
200 \\
\underline{180} \\
20
\end{array}
$$

25.
$$
\begin{array}{r}
.87\frac{4}{8} \doteq 0.87\frac{1}{2} \\
8\overline{).700} \\
\underline{640} \\
60 \\
\underline{56} \\
4
\end{array}
$$

26. $9 \div 10 = 0.90 = 90\%$

27. $2 \div 7 = 0.28\frac{4}{7} = 28\frac{4}{7}\%$

28. $1 \div 3 = 0.33\frac{1}{3} = 33\frac{1}{3}\%$

29. $3 \div 4 = 0.75 = 75\%$

30. $8 \text{ ft} \times 9 \text{ ft} \times 7.2 \text{ ft} = 518.4 \text{ ft}^3$

Lesson Test 24

1. $\dfrac{85}{100} = \dfrac{17}{20}$

2. $\dfrac{16}{100} = \dfrac{4}{25}$

3. $\dfrac{2}{10} = \dfrac{1}{5}$

4. $\dfrac{5}{1000} = \dfrac{1}{200}$

5. $2 \div 9 = 0.22\frac{2}{9} = 22\frac{2}{9}\%$

6. $4 \div 5 = 0.80 = 80\%$

7. $1 \div 2 = 0.50 = 50\%$

8. $2 \div 3 = 0.66\frac{2}{3} = 66\frac{2}{3}\%$

9. $0.75X = 7.5$
$\quad X = 7.5 \div 0.75 = 10$

10. $0.75(10) = 7.5$
$\quad 7.5 = 7.5$

11. $1.4X + 5 = 6.82$
$1.4X = 6.82 - 5$
$1.4X = 1.82$
$X = 1.82 \div 1.4 = 1.3$

12. $1.4(1.3) + 5 = 6.82$
$1.82 + 5 = 6.82$
$6.82 = 6.82$

13.
$$\begin{array}{r} 13.33\frac{1}{3} \\ 3\overline{)40.00} \\ \underline{3000} \\ 1000 \\ \underline{900} \\ 100 \\ \underline{90} \\ 10 \\ \underline{9} \\ 1 \end{array}$$

14.
$$\begin{array}{r} 2.43\frac{7}{11} \\ 11.\overline{)26.80} \\ \underline{2200} \\ 480 \\ \underline{440} \\ 40 \\ \underline{33} \\ 7 \end{array}$$

15. $1 + 5 + 9 = 15$
$15 \div 3 = 5$

16. $2 + 2 + 5 + 7 = 16$
$16 \div 4 = 4$

17. $75 \text{ lb} \times 1.30 = 97.5 \text{ lb}$

18. $45 \text{ in} + 31 \text{ in} + 45 \text{ in} + 31 \text{ in} = 152 \text{ in}$

19. $3.14(60 \text{ ft})^2 = 11,304 \text{ ft}^2$

20. $3.5 \text{ km} \times 1,000 \text{ m/km} = 3,500 \text{ m}$

Lesson Test 25

1. $8 + 13 + 17 + 17 + 19 + 19 + 19 = 112$
mean $= 112 \div 7 = 16$
median $= 17$
mode $= 19$

2. $70 + 80 + 80 + 90 + 100 = 420$
mean $= 420 \div 5 = 84$
median $= 80$
mode $= 80$

3. $8 + 8 + 9 + 10 + 15 = 50$
mean $= 50 \div 5 = 10$
median $= 9$
mode $= 8$

4. $3 + 3 + 6 + 7 + 11 = 30$
mean $= 30 \div 5 = 6$
median $= 6$
mode $= 3$

5. $\dfrac{53}{100}$

6. $\dfrac{44}{100} = \dfrac{11}{25}$

7. $23 \div 40 = 0.57\frac{1}{2} = 57\frac{1}{2}\%$

8. $4 \div 6 = 0.66\frac{2}{3} = 66\frac{2}{3}\%$

9. $9X + 3.4 = 7.9$
$9X = 7.9 - 3.4$
$9X = 4.5$
$X = 4.5 \div 9 = 0.5$

10. $9(.5) + 3.4 = 7.9$
$4.5 + 3.4 = 7.9$
$7.9 = 7.9$

11. $10 \times 10 = 100$

12. $3 \times 3 \times 3 = 27$

13. $1 \times 1 \times 1 \times 1 \times 1 \times 1 = 1$

14. $0.9 \text{ m} + 1 \text{ m} + 0.9 \text{ m} + 1 \text{ m} = 3.8 \text{ m}$

15. $0.9 \text{ m} \times 0.8 \text{ m} = 0.72 \text{ m}^2$

16. $9.7 \text{ ft} + 9.7 \text{ ft} + 9.7 \text{ ft} + 9.7 \text{ ft} = 38.8 \text{ ft}$

17. $9.7 \text{ ft} \times 9.7 \text{ ft} = 94.09 \text{ ft}^2$

18. $2 + 2 + 3 + 3 + 4 + 4 + 4 + 6 + 8 = 36$
average, or mean $= 36 \div 9 = 4$

19. most often, or mode $= 4$

20. middle, or median $= 4$

Lesson Test 26

1. $\dfrac{10}{1000} = \dfrac{1}{100}$

2. $\dfrac{300}{1000} = \dfrac{3}{10}$

3. $\dfrac{700}{1000} = \dfrac{7}{10}$

4. $\dfrac{4}{10} = \dfrac{2}{5}$

5. $\dfrac{4}{10} = \dfrac{2}{5}$

6. $\dfrac{7}{10}$

7. $3 + 7 + 8 + 8 + 9 = 35$
mean $= 35 \div 5 = 7$
median $= 8$
mode $= 8$

8. $9 + 9 + 12 + 13 + 17 = 60$
mean $= 60 \div 5 = 12$
median $= 12$
mode $= 9$

9. $\dfrac{5}{10} = \dfrac{1}{2}$

10. $\dfrac{92}{100} = \dfrac{23}{25}$

11. $\dfrac{9}{100}$

12. $\dfrac{825}{1000} = \dfrac{33}{40}$

13. $1 \div 3 = 0.33\dfrac{1}{3} = 33\dfrac{1}{3}\%$

14. $7 \div 8 = 0.87\dfrac{1}{2} = 87\dfrac{1}{2}\%$

15. $3.14(6.4 \text{ ft}) = 20 \text{ ft}$

16. $200 \times 0.16 = 32 \text{ people}$

17. 500 meters

18. $(8 + 4 - 6 + 3) \times 2 = 18$

Lesson Test 27

1. endpoint
2. length; width
3. ray
4. line segment
5. length; width
6. c
7. d
8. e
9. b
10. a
11. $\overrightarrow{XY}$
12. $\overleftrightarrow{QR}$ or $\overleftrightarrow{RQ}$
13. $1 + 1 + 1 + 2 + 5 = 10$
mean $= 10 \div 5 = 2$
14. median $= 1$

15. mode $= 1$
16. $2.05 \times 1.3 = 2.665$
17. $1.43 \div 0.11 = 13$
18. $1.75 + 2.25 = 4$
19. $\dfrac{1}{2}$
20. $\dfrac{50}{100} = \dfrac{1}{2}$
$8 \div 2 = 4$
$4 \times 1 = 4 \text{ chores done}$

Lesson Test 28

1. point
2. plane
3. plane
4. equal
5. congruent
6. similar
7. $\leftrightarrow$
8. $\sim$
9. ∞
10. $\rightarrow$
11. $\bullet$
12. $-$
13. $\square$
14. $\cong$
15. 50%
16. 25%
17. 75%
18. $33\dfrac{1}{3}\%$
19. 20%
20. $66\dfrac{2}{3}\%$

Lesson Test 29

1. rays
2. $\dfrac{1}{4}$
3. 90
4. 360
5. vertex
6. similar
7. congruent

8. length; width
9. ∠RGX or ∠XGR
10. $\dfrac{5}{10} = \dfrac{1}{2}$
11. $\dfrac{8}{100} = \dfrac{2}{25}$
12. $\dfrac{565}{1000} = \dfrac{113}{200}$
13. $\dfrac{\overset{2}{\cancel{4}}}{5} \times \dfrac{3}{\underset{5}{\cancel{10}}} = \dfrac{6}{25}$
14. $\dfrac{3}{24} + \dfrac{14}{24} = \dfrac{17}{24}$
15. $\dfrac{5}{8} - \dfrac{4}{8} = \dfrac{1}{8}$
16. $\dfrac{5}{150} = \dfrac{1}{30}$
17. 9 is most often, or mode
18. $16.34 + 37.2 + 45.5 + 61 = 160.04$ lb
    160.04 lb $\div 4 = 40.01$ lb

Lesson Test 30

1. 90º
2. acute
3. straight
4. obtuse
5. right
6. straight
7. obtuse
8. acute
9. 480%
10. 91%
11. 9%
12. $0.25 \times 300 = 75$
13. $1.50 \times 80 = 120$
14. $0.30 \times 75 = 22.5$
15. $\dfrac{2}{3} \div \dfrac{1}{8} = \dfrac{2}{3} \times \dfrac{8}{1} = \dfrac{16}{3} = 5\dfrac{1}{3}$
16. $\dfrac{4}{9} \div \dfrac{1}{3} = \dfrac{4}{\underset{3}{\cancel{9}}} \times \dfrac{\cancel{3}}{1} = \dfrac{4}{3} = 1\dfrac{1}{3}$
17. $15 \times 15 = 225$
18. $3.14(12.2 \text{ ft}) = 38.308$ ft

Unit Test IV

1. $\dfrac{52}{100} = \dfrac{13}{25}$
2. $\dfrac{8}{100} = \dfrac{2}{25}$
3. $\dfrac{3}{10}$
4. $\dfrac{575}{1000} = \dfrac{23}{40}$
5. $10 + 10 + 11 + 12 + 17 = 60$
   mean $= 60 \div 5 = 12$
   median $= 11$
   mode $= 10$
6. $8 + 13 + 17 + 17 + 19 + 19 + 19 = 112$
   mean $= 112 \div 7 = 16$
   median $= 17$
   mode $= 19$
7. $\dfrac{40}{280} = \dfrac{1}{7}$
8. $\dfrac{210}{280} = \dfrac{3}{4}$
9. $\dfrac{40 + 30}{280} = \dfrac{70}{280} = \dfrac{1}{4}$
10. $\overline{FG}$ or $\overline{GF}$
11. $\overleftrightarrow{RS}$ or $\overleftrightarrow{SR}$
12. $\overrightarrow{AW}$
13. ∠HJK or ∠KJH
14. ray
15. line segment
16. length; width
17. point
18. plane
19. congruent
20. equal
21. similar
22. vertex
23. 360
24. 90º
25. acute
26. straight
27. obtuse
28. c
29. e
30. f
31. b
32. d

33. g
34. a

Final Test

1. $1 \times 1 \times 1 \times 1 \times 1 \times 1 = 1$
2. $8 \times 8 = 64$
3. $10 \times 10 \times 10 = 1000$
4. 5,271.349
5.
```
   6  14
   7. 5 2
 - 1. 8 5
   5. 6 7
```
6.
```
    6.0
  +5.28
  11.28
```
7.
```
   1  9  13
 3 2. 0 4 1
 -0. 5 9 6
 31. 4 4 5
```
8.
```
    2.4 9
  ×   .6
    2 5
  1 2 4 4
  1.4 9 4
```
9.
```
    1.7
  × 3
    2
   3 1
   5.1
```
10.
```
    .0 0 4
  ×.0 5
  .0 0 0 2 0
```
11. 1,300,000 cm
12. 0.25 g
13. 0.05
14. 0.65
15. $\frac{25}{100} = \frac{1}{4}$
16. $\frac{32}{100} = \frac{8}{25}$
17. $0.8 = 80\%$
18. $5 \div 6 = 0.83\frac{1}{3} = 83\frac{1}{3}\%$
19. $\frac{400}{100} + \frac{60}{100} = \frac{460}{100} = 4.60 = 460\%$

20. $\frac{78}{100} = \frac{39}{50}$
21. $\frac{3}{100}$
22.
```
        3.325 ≈ 3.33
  4 13.300
    12 00
     130
     120
      10
       8
       2
```
23.
```
       .654 ≈ 0.65
  7 4.580
    4200
     380
     350
      30
      28
       2
```
24.
```
        65.66 ≈ 65.6̅
  .6 39.400
     36000
      3400
      3000
       400
       360
        40
        36
         4
```
25.
```
        .733 ≈ 0.7̅3̅
  .03 .02200
       2100
        100
         90
         10
          9
          1
```
26.
```
       .81 9
          ──
          11
  11 9.00
     880
      20
      11
       9
```

27.

$$9\overline{)5.00} \quad .55\frac{5}{9}$$

$$\underline{450}$$
$$50$$
$$\underline{45}$$
$$5$$

28. $3.2X + 0.07 = 4.55$

$\quad 3.2X = 4.55 - 0.07$

$\quad 3.2X = 4.48$

$\quad\quad X = 4.48 \div 3.2$

$\quad\quad X = 1.4$

29. $3.2(1.4) + 0.07 = 4.55$

$\quad 4.48 + 0.07 = 4.55$

$\quad\quad 4.55 = 4.55$

30. plane

31. line segment

32. point

33. length; width

34. ray

35. obtuse

36. acute

37. similar

38. 360

39. 90°

40. straight

41. congruent

42. $A = 3.14(3 \text{ ft})^2 = 28.26 \text{ ft}^2$

$\quad C = 3.14(6 \text{ ft}) = 18.84 \text{ ft}$

43. $\$3.50 + \$5 + \$5 + \$7 + \$8 = \$28.50$

$\quad$ mean $= \$28.50 \div 5 = \$5.70$

$\quad$ median $= \$5$

$\quad$ mode $= \$5$

44. $\$45.60 \times 1.14 \approx \$51.98$

45. $\dfrac{1}{758}$

Word Problems 6

1. $3.45 + $1.99 + $6.59 + $12.98 = $25.01
$25.01 - $2.50 = $22.51
$50.00 - $22.51 = $27.49 is money left

2. Use a drawing to show their travels.
The distances don't have to be to scale.

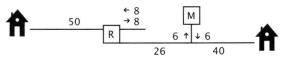

50 mi + 26 mi + 40 mi = 116 miles from
house to house
50 mi + 8 mi + 8 mi = 66 miles past
restaurant and back
66 mi + 26 mi = 92 miles to turnoff
92 mi + 6 mi + 6 mi = 104 miles back
to main route
104 mi + 40 mi = 144 miles total driven

3. 3.5" - 0.6" = 2.9"
2.9" + 8.3" = 11.2"
11.2" - 4.2" = 7" remaining

Word Problems 12

All money is rounded to hundredths at each step

1. $8 \times $8.40 = $67.20
$7 \times $5.99 = $41.93
$67.20 + $41.93 = $109.13 discountable
$109.13 \times 0.10 \approx $10.91 (rounded)
$109.13 - $10.91 = $98.22
for discounted yarn
$2 \times $2.50 = $5.00 for un-discounted
$98.22 + $5.00 = $103.22 cost of yarn
$103.22 \times 1.13 = $116.64
with tax and shipping (5% + 8% = 13%)

2. $5.99 \times 5 = $29.95
$0.10 \times $29.95 = 2.995 rounds to $3.00
$29.95 - $3.00 = $26.95 discounted price
$26.95 \times 1.13 = $30.45
with tax and shipping
   You could also figure the total cost of one
skein and multiply by 5. Rounding may
give a slightly different answer.

3. $2.50 \times 1.13 \approx 2.83
for one skein (rounded)
$2.83 \times 1.5 \approx $4.25 (rounded)
   You could have found the basic cost of
1.5 skeins first, and then figured tax and
shipping. Your answer may be slightly
different because of rounding.

Word Problems 18

1. $\frac{1}{4} + \frac{1}{2} + \frac{3}{4} + \frac{1}{4} =$
$\frac{1}{4} + \frac{2}{4} + \frac{3}{4} + \frac{1}{4} =$
$\frac{7}{4}$ or $1\frac{3}{4}$ pizza left over
$\frac{7}{4} \div \frac{1}{4} = \frac{7}{4} \times \frac{4}{1} = 7$ boys

2. 5.3" ÷ 10 = 0.53"
water from average snow
4.1" ÷ 5 = 0.82" water from wet snow

If you are unsure whether to multiply or
divide, check your answer to see if it
makes sense. The actual snowfall in each
case was less than what was needed to
make an inch of water, so division yields
a sensible answer.
0.53" + 0.82" + 1.5" = 2.85"
rounds to 2.9" of water

3. (2)3.14 × 3' = 18.84'
circumference of small garden
2 × 3' = 6' diameter of small garden
3.14 × 12' = 37.68' circumference
of garden with doubled diameter
37.68' - 18.84' = 18.84'
additional edging needed

In real life this would probably be
rounded to 19 or 20 feet.

4. $3.14 \times (3 \text{ ft})^2 = 28.26 \text{ ft}^2$ area
of small garden
28 ft^2 rounded
$3.14 \times (6 \text{ ft})^2 = 113.04 \text{ ft}^2$ area
of large garden
113 ft^2 rounded
$113 \text{ ft}^2 \div 28 \text{ ft}^2 / \text{pack.} = 4.04$ or
4 seed packets (rounded)
When you have learned how to divide a
decimal by a decimal, try this again using
the unrounded areas.

Word Problems 24

1. $8 \text{ in} \times 12 \text{ in} = 96 \text{ in}^2$ area of paper
$3.14(3 \text{ in})^2 = 28.26^2 \text{ in}$ area of one circle
$3.14(2.5 \text{ in})^2 = 19.625 \text{ in}^2$ or 19.63 in^2
rounded area of other circle
$28.26 \text{ in}^2 + 19.63 \text{ in}^2 = 47.89 \text{ in}^2$
used for circles
$96 \text{ in}^2 - 47.89 \text{ in}^2 = 48.11 \text{ in}^2$ left over
Look at the drawing to see why it is
not possible to cut another circle with a
3" radius, even though there seems to
be enough area.

One circle has a diameter of 6", which
leaves 2" distance to the edge of the
paper.

The other circle has a diameter of 5",
which leaves a 3" distance to the edge
of the paper.

Neither space is enough for another
circle with a 3" radius (6" diameter).

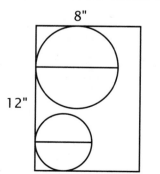

2. $\$6.50 \times 0.10 = \$.65$ is 10%
of his hourly income
$\$6.50 - 0.65 = \$5.85$
hourly amount available to spend
$\$25.35 + \$50.69 + \$85.96 =$
$\$162$ total needed
$\$162 \div \$5.85 = 27.69...$
rounds to 28 hours

3. $0.3 \times T = 300$ blue toys
$T = 300 \div 0.3$
$T = 1000$ toys in all
$\dfrac{1}{4} \times 1000 = 250$ green toys
$300 + 250 = 550$ blue or green toys
$1000 - 550 = 450$ remaining toys
$\dfrac{1}{2} \times 450 = 225$ red toys
$\dfrac{1}{2} \times 450 = 225$ yellow toys

You could use decimals or fractions
for this problem.

Symbols & Tables

U. S. CUSTOMARY MEASUREMENT

3 teaspoons (tsp) = 1 tablespoon (Tbsp)

2 pints (pt) = 1 quart (qt)

8 pints = 1 gallon (gal)

4 quarts = 1 gallon

12 inches (in) = 1 foot (ft)

3 feet = 1 yard (yd)

5,280 feet = 1 mile (mi)

16 ounces (oz) = 1 pound (lb)

2,000 pounds = 1 ton

METRIC MEASUREMENT

1,000 millimeters (mm) = 1 meter (m)

100 centimeters (cm) = 1 meter

10 decimeters (dm) = 1 meter

10 meters = 1 dekameter (dam)

100 meters = 1 hectometer (hm)

1,000 meters = 1 kilometer (km)

Replace meter with liter or gram for liquid and weight equivalents.

METRIC – U. S. EQUIVALENTS

1 meter ($\approx$ 39.37 in) $\approx$ 1.09 yd (36 in)

1 centimeter $\approx$ 0.4 inch

1 liter $\approx$ 1.06 quarts

1 kilogram $\approx$ 2.2 pounds

1 kilometer $\approx$ 0.6 mile

1 inch $\approx$ 2.5 cm or 1 inch = 2.54 cm

1 ounce $\approx$ 28 grams

SYMBOLS

=	equals
$\sim$	similar
$\cong$	congruent
$\approx$	approximately equal to
<	less than
>	greater than
%	percent
π	pi (22/7 or 3.14)
r^2	r squared or $r \cdot r$
'	foot
"	inch
∞	infinity
$\cdot$	point
$\leftrightarrow$	line
$\rightarrow$	ray
$\angle$	angle
▱	plane
⦜	right angle
| |	absolute value
$.\overline{3}$	decimal repeat

PERIMETER

rectangle, square, triangle, parallelogram – add the lengths of all the sides

CIRCUMFERENCE

circle $C = 2\pi r$

AREA

rectangle, square, parallelogram

$A = bh$ (base times height)

triangle

$A = \dfrac{bh}{2}$ or $\dfrac{1}{2}bh$

circle

$A = \pi r^2$

VOLUME

rectangular solid, cube

$V = Bh$ (area of base times height)

Glossary

A

Absolute Value - the distance of a number from zero; always positive

Acute angle - an angle with a measure greater than 0° and less than 90°

Algebraic expression - a combination of one or more numbers, symbols representing numbers, and operation symbols

Angle - a geometric figure formed by two rays joined at their origins

Area - the amount of space occupied by a two-dimensional figure

Average - usually refers to the mean, which is the result of adding a series of numbers and dividing by the number of items in the series

B–D

Base - the side (in a two dimensional shape) or surface (in a three dimensional shape) which is perpendicular to the height

Base 10 - a number system where the number words or symbols are multiplied by multiples of ten. (Multiples of ten are numbers with a base of 10.)

Box Plot - a graph which shows spread and variability in a set of data by focusing attention on the median and interquartile range

Centi - in the metric system, the Latin prefix representing one-hundredth

Circle - a figure where all the points are an equal distance from a center point

Circumference - the distance around a circle. It corresponds to perimeter.

Cluster - a group of data with similar values

Coefficient - a factor within a term, usually we use this word to refer specifically to the number that a variable is multiplied by

Commutative property (of addition) - the order of addends may be changed without changing the sum

Commutative property (of multiplication) - the order of factors may be changed without changing the product

Congruent - having the same shape and size

Conversion factor - a unit rate used as a factor to convert units; see also unit multiplier

Coordinates - a pair of numbers used to designate a specific point on a two-dimensional graph

Counting numbers - whole numbers greater than zero

Cube - a three-dimensional figure with each side the same length

Deci - in the metric system, the Latin prefix representing one-tenth

Decimal or decimal fraction - a fraction written on one line by using a decimal point and place value

Decimal notation - a system of writing numerals where the position (place) of a symbol corresponds to a multiple of ten, as in our Arabic numeral system

Decimal point - the dot used in decimal numbers between the units place and the tenths place

Decimal system - a number system where the number words or symbols are multiplied by multiples of ten (see base 10).

Deka or deca - in the metric system, the Greek prefix representing 10

Denominator - the bottom number in a fraction. It tells how many total parts there are in the whole.

Diameter - The length of a line segment that passes through the center of a circle and that touches both edges of a circle. It is twice the length of the radius.

Distributive Property (of multiplication)- a factor may be multiplied by each term of an expression that is a factor

Dividend - the number being divided in a division operation

Divisor - the number that is being divided by in a division operation

Dot plot - a graph that uses dots or other symbols to show how many times each piece of data occurs

E–G

Endpoint - one of the starting or stopping points of a line segment, or the starting point of a ray

Equal - having the same numerical value

Equation - a "number sentence" in which the value of one side is equal tot he value of the other side; two expressions or numbers that are equivalent and joined by an equal sign

Equivalent - having the same value

Estimation - a process used to get an approximate value of an answer

Even number - a number that ends in 0, 2, 4, 6, or 8; multiples of two.

Expanded notation - a way of writing numbers in which each amount is explicitly multiplied by its place value

Exponent - a small number written to the upper right of another number. It tells how many times the first number (the base) is to be used as a factor.

Exponential notation - a way of writing numbers in which each amount is explicitly multiplied by its place value and in which each place value is indicated by 10 with an exponent

Factors - the numbers multiplied in a multiplication operation

Fraction - a number shown by writing one numeral over another. The top number (numerator) is divided by the bottom number (denominator).

Geometry - the study of shapes and their sizes and relative positions. Plane geometry studies one- and two-dimensional figures. Solid geometry studies three-dimensional figures.

Gram - the base unit of weight or mass in the metric system

Greatest Common Factor - the largest number that is a factor of two or more given numbers

Greatest common factor (GCF) - the largest whole number factor that will go evenly into two or more given numbers

H–L

Hecto - in the metric system, the Greek prefix representing one hundred

Height - the length of a line from the top to the bottom of a shape that forms a right angle with the base

Histogram - a type of bar graph

Histogram - a bar graph in which each bar represents a range or category of data

Improper fraction - a fraction with a numerator larger than its denominator

Interquartile range (IQR) - the range of the middle 50% of a data set, used to measure the spread of a data set.

Inverse - opposite. Multiplication and division are inverse operations, and addition and subtraction are inverse operations.

Infinity - a quantity greater than any number

Kilo - in the metric system, the Greek prefix representing one thousand

Least Common Multiple (LCM) - the smallest whole number that is a multiple of a given set of numbers

Line - in plane geometry, a geometric figure that has infinite length and no width, and is perfectly straight

Line segment - a measurable piece of a line. It has two endpoints.

Liter or litre - the base unit of volume or capacity in the metric system

M–O

Mean - the result of adding a series of numbers and dividing by the number of items in the series.

Mean absolute deviation (MAD) - a measure of the variation in a data set

Median - the middle number in a set of numbers, when they are arranged from smallest to largest

Meter or metre - the base unit of linear measure in the metric system

Metric system - a system of measurement based on 10

Milli - in the metric system, the Latin prefix representing one-thousandth

Mixed number - a number made up of a whole number and a fraction

Mode - the number that occurs most often in a set

Negative number - a number less than zero

Net - a diagram showing a three-dimensional figure flattened to a two-dimensional shape

Number line - a straight line with marks indicating a scale of numbers

Numerator - the top number in a fraction. It designates a number of the parts of a whole chosen.

Obtuse angle - an angle with a measure greater than 90° and less than 180°

Origin - the starting point of a ray; also the point where the veritical and horizontal axes cross in a graph [(has coordinates of (0, 0)]

P–R

Parallel - two lines in the same plane that never touch each other

Parallelogram - a shape with four sides formed by two sets of parallel lines.

Peak - a high point of the data on a graph

Per cent - "per hundredth" - a way of writing fractions that have a denominator of 100. The symbol is %.

Perimeter - the distance around a two-dimensional figure

Perpendicular - refers to two lines that form a right angle where they meet, two lines perpendicular to the same line are parallel.

Pi - he Greek letter π. It is used to represent the ratio of the diameter of a circle to its circumference, and it is close in value to the numbers 22/7 and 3.14

Pie graph - a circular graph used to show what percent of different parts are in a whole

Place value - the position of a number that tells what value it is assigned

Place–value notation - a way of writing numbers used to emphasize the place value of each part by writing each place as a separate addend in order

Plane - a flat, two-dimensional figure that extends ininitely in all directions.

Plane geometry - study of one- and two-dimensional figures

Point - the smallest figure in plane geometry. It has no length or width.

Power - another name for an exponent. 2^3 can be read "two to the power of three."

Probability - a numerical expression of how likely something is to happen or a statement is to be true

Product - the result to a multiplication operation

Proper fraction - a fraction with a numerator smaller than its denominator

Quadrants - four sections of a two-dimensional graph created by the x-axis and y-axis

Quotient - the result of a division operation

Radius - the distance from the center of a circle to its edge

Rate - a ratio where the units for the two parts are different

Ratio - the relationship between two numbers which can be written in the form a:b or a/b

Rational number - a number that can be expressed as a fraction or ratio

Ray - a geometric figure that starts at a definite point and extends infinitely in only one direction

Reciprocal - the number which may be multiplied by another number to produce one; when written as fractions reciprocals appear to have the numerator and denominator "flipped"

Rectangle - a shape with four "square corners" or right angles

Rectangular solid - a three-dimensional shape with each side or face shaped like a rectangle

Reducing - dividing the numerator and denominator of a fraction by a common factor. The resulting fraction is equivalent but in "lower terms."

Repeating decimal - a decimal with a pattern

Right angle - 90° angle (square corner)

Rounding - replacing a number with a more convenient nearby number, for example writing a number as its closest multiple of ten in order to estimate.

Rule of four - a method for finding the same, or common, denominator of two fractions

S–U

Similar - having the same shape but not the same size

Solid geometry - study of three-dimensional figures

Square - a rectangle with all four sides the same length

Statistics - the branch of mathematics that collects and organizes data to find patterns and make predictions or draw conclusions.

Straight angle - an angle with a measure of 180°

Surface area - the sum of the areas of all the surfaces of a three-dimensional figure

Symmetrical - term used to describe a data set if the shape of its graph is similar on both sides of a line drawn through the median.

Term - one part of an expression that is being added to or subtracted from another part of the expression

Triangle - a closed shape with three sides

Unit multiplier - a rate used as a factor to convert units; conversion factor

Unit rate - a rate where one of the values is one

Unknown - a number whose value we do not know. It is usually represented by some letter of the alphabet.

V–Z

Variable - a letter representing an unspecified number from a specified group of numbers

Variability - the extent that a data set varies from its mean

Vertex - the origin of the rays of an angle or the point of the angle

Volume - the space occupied by a three-dimensional shape

Whole numbers - zero and all positive numbers that are a whole and do not need to be written as fraction

X-axis - the horizontal axis of a two-dimensional graph

Y-axis - the vertical axis of a two-dimensional graph

Master Index for General Math

This index lists the levels at which main topics are presented in the instruction manuals for *Primer* through *Zeta*. For more detail, see the description of each level at MathUSee.com. (Many of these topics are also reviewed in subsequent student books.)

Zeta Index